SCHAUM'S A–Z

MATHEMATICS

SCHAUM'S A-Z

MATHEMATICS

JOHN BERRY, TED GRAHAM, JENNY SHARP, AND ELIZABETH BERRY

SCHAUM'S A-Z SERIES

McGraw·Hill

New York   Chicago   San Francisco   Lisbon   London   Madrid   Mexico City
Milan   New Delhi   San Juan   Seoul   Singapore   Sydney   Toronto

Originally published by Hodder & Stoughton Educational, a division of Hodder Headline Plc, 338 Euston Road, London NW1 3BH.

2 3 4 5 6 7 8 9 0   2 1 0 9 8 7 6 5 4 3

ISBN 0-07-141936-7

McGraw-Hill books are available at special quantity discounts to use as premiums and sales promotions, or for use in corporate training programs. For more information, please write to the Director of Special Sales, Professional Publishing, McGraw-Hill, Two Penn Plaza, New York, NY 10121-2298. Or contact your local bookstore.

HOW TO USE THIS BOOK

The *Schaum's A–Z* is an alphabetical textbook designed for ease of use. Each entry begins with a short definition or explanation. Many entries have worked examples showing typical short examination type questions.

In writing each entry we have kept three important questions in mind. "What does the entry mean? Why do I need to know it? How is it used?" This follows the familiar view of mathematics knowledge consisting of "concept, context and skill."

The *Schaum's A–Z* is not designed to replace your textbook or teacher! We hope that you will see it as a helpful complementary part of your studies. It is designed for you to use as you study new topics in mathematics to provide a dictionary of words and techniques that you become familiar with. When you meet a new topic in mathematics check its meaning in the *Schaum's A–Z* to get an initial overview. Return to the handbook from time to time to review keywords, phrases or ideas and have it by your side during your exam period.

We hope that the *Schaum's A–Z* proves an invaluable resource, fully relevant from the first day of your mathematics, mechanics, statistics or decision and discrete mathematics course and then on after school, college or university.

John Berry, Ted Graham, Jenny Sharp and Elizabeth Berry

ACKNOWLEDGMENTS

Researching, writing and editing a book of this size requires teamwork and hard work. It also requires the occasional willingness to sacrifice technical accuracy in favor of clarity. A considerable amount of our time went into checking the entries, but, if any mistakes have slipped through, the authors accept full responsibility.

The authors are pleased to acknowledge the support of our colleagues in the Centre for Teaching Mathematics, Wendy, Stuart, Roger, Karen and Raymond for helping to clarify our understanding of some of the entries! In particular we thank Karen Eccles for helping to prepare the manuscript. The other people to be thanked are Tim Sipka of Alma College, Michigan and the team at Hodder Headline.

John Berry, Ted Graham, Jenny Sharp and Elizabeth Berry
The Centre for Teaching Mathematics
University of Plymouth

The authors and publishers would like to thank the following for permission to reproduce material in this volume:

Dr Ted Graham for the photo on page 61 and Pictor International Ltd., London for the photos on pages 28, 78, 80 and 92.

absolute and relative error: absolute error is the actual error that exists when an estimate or approximation is made. For example, if a rod is measured as having a length of 199.5 cm when its length is in fact exactly 200 cm, then the absolute error is 0.5 cm.

The relative error is given by the absolute error divided by the true value and is often given as a percentage. For the example above the relative error is:

$$\frac{0.5}{200} = 0.0025 = 0.25\%$$

absolute value: the absolute value of a number, x, is written as $|x|$, the modulus of x. The modulus of a number is the size of the number, that is the number without its sign. For example $|5| = 5$, but $|-7| = 7$. This means that $|x|$ is never negative. The graph below shows $y = |x|$.

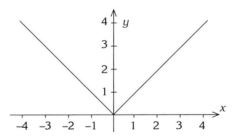

acceleration: defined as the rate of change of velocity. It is a vector quantity, having both magnitude and direction.

If an object is traveling in a straight line with constant speed, its velocity is constant and its acceleration is zero.

If an object is traveling along a curve with constant speed then the acceleration is variable because the direction of motion, and hence the velocity, are changing. So constant speed does not always mean zero acceleration.

The table shows the various situations that can occur.

velocity		acceleration
magnitude (speed)	direction	
constant	constant	zero
variable	constant	nonzero
constant	variable	nonzero
variable	variable	nonzero

1

We often denote the acceleration by the symbol **a**; if the velocity of the object is **v** then:

$$\mathbf{a} = \frac{d\mathbf{v}}{dt}$$

For an object moving in a straight line with position $x(t)$ and velocity $v(t)$ then:

$$a = \frac{dv}{dt} \qquad \text{or} \qquad a = v\frac{dv}{dx} \qquad \text{or} \qquad a = \frac{d^2x}{dt^2}$$

(See also *angular acceleration, constant acceleration equations*.)

acceptance region: if the *test statistic* falls in the acceptance region the *null hypothesis*, H_0, is accepted. The acceptance region is defined by the *critical values* of the test and can be illustrated diagramatically.

For a *two-tailed* test:

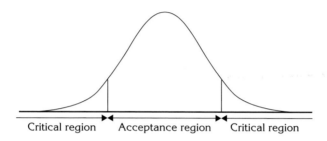

Critical region   Acceptance region   Critical region

and a *one-tailed* test:

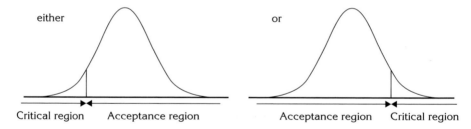

either   or

Critical region   Acceptance region   Acceptance region   Critical region

accuracy: when a number is quoted to a certain degree of accuracy we can place certain bounds on the actual value of the number. For example if $x = 16.7$ to 1 decimal place, then $16.65 \le x < 16.75$. (See also *absolute and relative errors*.)

activity network: see *precedence network*.

activity on arc network: a precedence network in which the *arcs* represent the activities, while the *vertices* represent *events*. (See *precedence network*.)

activity on vertex network: a precedence network in which the activities are represented by the *vertices* of a *network*, and *edges* show the order of precedence. (See *precedence network*.)

acute: an acute angle is an angle smaller than 90°. In an acute triangle all the angles are less than 90°.

addition and subtraction of complex numbers: when complex numbers are added the real parts of the numbers must be added together and the imaginary parts added together separately. The addition and subtraction of complex numbers can be treated geometrically as shown below.

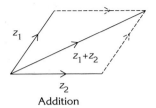
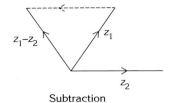

Addition   Subtraction

Example:

Three complex numbers are

$z_1 = 4 + 2i,$ $z_2 = -5 + 6i$ $z_3 = 4 - 8i.$

Find

(a) $z_1 + z_2$ (b) $z_1 - z_3$ (c) $z_1 + z_2 - z_3$

Solution:

(a)

$$z_1 + z_2 = (4 + 2i) + (-5 + 6i)$$
$$= (4 + (-5)) + (2 + 6)i$$
$$= -1 + 8i$$

(b)

$$z_1 - z_3 = (4 + 2i) - (4 - 8i)$$
$$= (4 - 4) + (2 + 8)i$$
$$= 10i$$

(c)

$$z_1 + z_2 - z_3 = (4 + 2i) + (-5 + 6i) - (4 - 8i)$$
$$= (4 + (-5) - 4) + (2 + 6 - (-8))i$$
$$= -5 + 16i$$

addition of vectors: see *vectors*.

adjacency matrix: a matrix representing the *vertices* and *edges* of a *graph*. Each row and column of the matrix represents a *vertex* of a graph and the numbers give the number of edges joining each pair or vertices.

	A	B	C	D
A	0	2	1	1
B	2	0	0	1
C	1	0	0	1
D	1	1	1	2

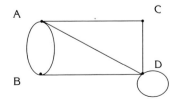

algorithm: a systematic process for finding a solution to a problem.

alternating path: a path in a *bipartite graph* that joins *vertices* from one subset to vertices in the other, in such a way that alternate *edges* only are in the initial matching, and initial and final vertices are not incident with an edge in the matching.

alternative hypothesis (H_1): this is the hypothesis that is accepted if the *null hypothesis* H_0 is rejected when performing a hypothesis test. The alternative hypothesis depends on the type of test being performed: *one-tailed* or *two-tailed*. A one-tailed hypothesis test considers strictly an increase or a decrease but not both, whereas a two-tailed hypothesis test considers any change in the parameter.

For example, consider a sample of peas from a species which was known to have a mean mass of 0.1 g. We wish to test if the mean mass of the peas in the sample differs from 0.1 g. Here the null hypothesis is $H_0 : \mu = 0.1$ and the alternative hypothesis is $H_1 : \mu \neq 0.1$. This is a two-tailed test.

If we wished to test if the mean mass of the peas in the sample had increased, the alternative hypothesis would be $H_1 : \mu > 0.1$. If testing for a decrease then it would be $H_1 : \mu < 0.1$. These would be one-tailed hypotheses.

altitude of a triangle: an altitude of a triangle is a line that is perpendicular to one side and passes through the opposite vertex.

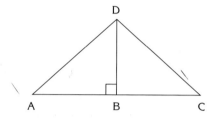

In the diagram BD is an altitude of the triangle ACD.

amplitude: the amplitude of oscillating motion of an object is the largest displacement from the equilibrium position of that object.

In the figure the object is moving up and down on the end of a spring.

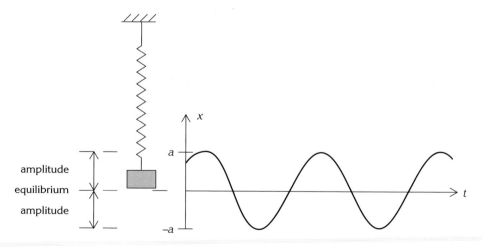

For an ideal situation as the object moves, the displacement from equilibrium x will form a trigonometric curve between limits $x = -a$ and $x = +a$. This is called *simple harmonic motion*.

The equation of the curve is:

$$x = a \cos(\omega t + \varepsilon)$$

and a is the amplitude of the motion.

In a real system the oscillation will gradually reduce in size. We say that the oscillation is *damped*.

For such a system the graph of x against t is shown below.

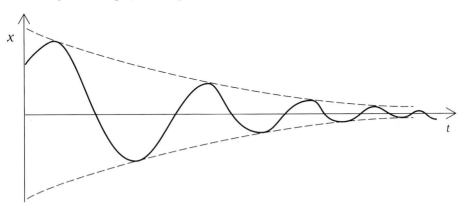

The bounding curve describes the amplitude, which is a decreasing function of time.

angle between a line and a plane: the angle between the line and its projection onto the plane. In the figure below, the line AB is the projection of the line AC onto the plane and is such that the line BC is perpendicular to the plane.

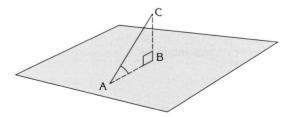

angle between two planes: the angle between lines in each plane that are both perpendicular to the line of intersection of the planes.

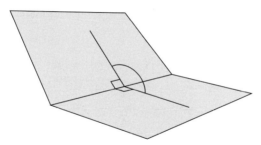

Example:

The points A, B, C and D have coordinates (0, 0, 2), (6, 0, 0), (0, 6, 0) and (0, 0, 8), respectively. Find the angle between the planes ABC and BCD.

Solution:

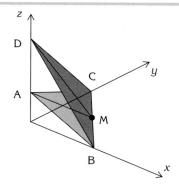

The diagram shows the planes and the midpoint of the line BC. The required angle is the angle between the lines AM and DM. The coordinates of M are:

$$\left(\frac{6 + 0}{2}, \frac{0 + 6}{2}, \frac{0 + 0}{2} \right) = (3, 3, 0)$$

The length of AM is:

$$\sqrt{(3 - 0)^2 + (3 - 0)^2 + (0 - 2)^2} = \sqrt{22}$$

The length of DM is:

$$\sqrt{(3 - 0)^2 + (3 - 0)^2 + (0 - 8)^2} = \sqrt{82}$$

The length of AD is 6.

Using the cosine rule in the form

$$\cos \theta = \frac{b^2 + c^2 - a^2}{2bc}, \text{ with angle AMD} = \theta, b = \sqrt{22}, c = \sqrt{82}$$

gives:

$$\cos \theta = \frac{22 + 82 - 6^2}{2 \times \sqrt{22} \times \sqrt{82}}$$

$$= 0.8005 \quad \text{so} \quad \theta = 36.82°$$

angle bisector: an angle bisector cuts an angle into two equal parts, as shown in the diagram below.

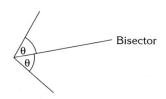

Bisector

angle between two lines: the angle between the two lines $y = m_1x + c_1$ and $y = m_2x + c_2$ is θ, where

$$\tan \theta = \frac{m_1 - m_2}{1 + m_1m_2}$$

Note that if $m_1 = m_2$ the lines are parallel and that if $m_1 \times m_2 = -1$ the lines are perpendicular.

Example:

Find the angle between the lines $y = 4x - 3$ and $y = 5x + 1$.

Solution:

The lines have gradients 4 and 5, so the angle θ between the lines is given by:

$$\tan \theta = \frac{5 - 4}{1 + 5 \times 4}$$

$$= \frac{1}{21}$$

$$\theta = 2.73°$$

angular speed: suppose that a point P moves in a plane so that its position in polar coordinates is (r, θ) then the angular speed of the point is the rate of change of the angle θ.

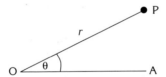

In the figure the point O and the line OA are fixed. In symbols the angular speed is often denoted by ω, so that

$$\omega = \frac{d\theta}{dt}$$

The units of angular speed are usually radians per second; however in many applications it makes more sense to use revolutions per minute (rev/min).

Example:

A compact disc on a hi-fi system rotates at 33⅓ rev/min. Convert this angular speed to radians per second.

Solution:

1 revolution = 2π radians and 1 minute= 60 seconds.

$$\text{So } 33\tfrac{1}{3} \text{ revolutions per minute } = \frac{33.33 \times 2\pi}{60}$$

$$= 3.49 \text{ radians per second}$$

(See also *circular motion*.)

angular acceleration: defined as the rate of change of angular velocity.

annulus: an annulus is the region between two concentric circles. This is the shaded region shown in the diagram below.

arc: an arc is a continuous part of a curve. For example a part of a circle is usually referred to as an arc. The diagram below shows a circle split into two arcs by the points A and B.

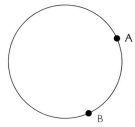

arc (or edge): a line connecting two *vertices* in a graph. For a *directed network* each *edge* has a direction of "flow" associated with it.

arccos: a term used for the inverse of the *cosine* function, more usually written cos⁻¹; see *cos⁻¹* for the definition.

arc length: the term arc length is used to describe the distance along a curve between two specified points. The length of an arc can be calculated easily for an arc that is part of a circle. The figure shows a circle and the arc length between the points A and B can be calculated using one of the two formulas.

$$\text{arc length} = \frac{2\pi r \theta}{360} \quad \text{if } \theta \text{ is in degrees.}$$

$$\text{arc length} = r\theta \quad \text{if } \theta \text{ is in radians.}$$

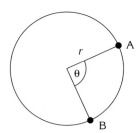

arcsin: a term used for the inverse of the *sine* function. It is more common to use sin⁻¹, and readers should consult the *sin⁻¹* entry.

arctan: term used for the inverse of the *tangent* function. It is more common to use tan^{-1}, and readers should consult the *tan^{-1}* entry.

area: an area is a region enclosed within a boundary. For example, the area of a rectangle of sides length a and b is ab. The area between the graphs of two functions can be evaluated using integration.

Example:

Find the area enclosed by the two curves $y = x^2$ and $y = \sqrt{x}$.

Solution:

First we sketch the functions to see where the area lies in the x–y plane.

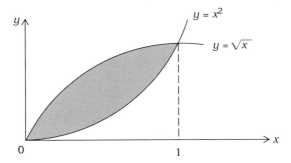

The graphs intersect at $x = 0$ and $x = 1$ and the area required is shown shaded in the diagram.

$$\text{Area} = \int_0^1 \sqrt{x} - x^2 \, dx$$

$$= \left[\frac{2}{3} x^{3/2} - \frac{1}{3} x^3 \right]_0^1 = \frac{2}{3} - \frac{1}{3} = \frac{1}{3}$$

The area between the curves is $\dfrac{1}{3}$.

argand diagram: an argand diagram is used to give a geometrical interpretation to a complex number. The diagram below shows how the complex number $z = a + bi$ can be represented.

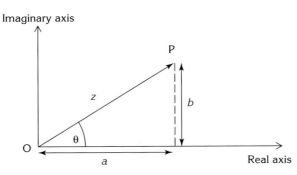

The complex number z is represented by the point P or the line OP. The modulus of the complex number z is written as $|z|$ and is the length of OP. It is calculated using $|z| = \sqrt{(a^2 + b^2)}$. The argument of the number z, written as arg z, is the angle between the line OP and the real axis, shown as θ on the diagram. It can be found using

$$\text{arg } z = \tan^{-1}\left(\frac{b}{a}\right) \text{ where } -\pi < \text{arg } (z) \leq \pi$$

If the modulus and argument of a complex number are r and θ, respectively, the number can be expressed in the form $z = a + bi$. The diagram shows how a and b can be expressed in terms of r and θ, so that $z = r \cos \theta + r \sin \theta i$.

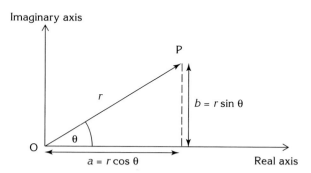

argument of a complex number: this is the angle between the complex number, when it is represented as a line, and the real axis. The argument of the complex number $z = a + bi$ is written as arg(z) and is the angle θ in the triangle above.

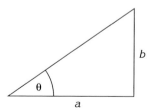

For the figure shown above:

$$\text{arg } (z) = \tan^{-1}\left(\frac{b}{a}\right) \qquad \text{where} \qquad -\pi < \text{arg } (z) \leq \pi$$

arithmetic mean: the arithmetic mean or simply the *mean* of a set of numbers $(x_1, x_2, \ldots x_n)$ is denoted by $\bar{x}$ (pronounced x bar) and is a measure of the central location or *average* of the set of numbers. It is given by the formula:

$$\bar{x} = \frac{1}{n}\sum_{i=1}^{n} x_i$$

The arithmetic mean is a good *measure of location for symmetric data*. Most calculators will calculate the mean of a set of data.

Example 1:

Find the mean of the following numbers: 3, 2, 5, 6, 4, 2, 3, 6, 7, 1, 2, 4.

Solution:

$$\bar{x} \;=\; \frac{1}{n}\sum_{i=1}^{n} x_i$$

$$=\; \frac{1}{12}(3 + 2 + 5 + 6 + 4 + 2 + 3 + 6 + 7 + 1 + 2 + 4)$$

$$=\; \frac{45}{12} = 3.75$$

For a *discrete frequency distribution* where the values $(x_1, x_2, \ldots x_n)$ have corresponding frequencies of $(f_1, f_2, \ldots f_n)$, the arithmetic mean is given by

$$\bar{x} \;=\; \frac{\displaystyle\sum_{i=1}^{n} f_i x_i}{\displaystyle\sum_{i=1}^{n} f_i}$$

Example 2:

Find the mean of the following data.

x	1	2	3	4	5	6	7
frequency f	6	4	6	7	2	8	3

Solution:

Compile a *frequency table*:

x	f	fx
1	6	6
2	4	8
3	6	18
4	7	28
5	2	10
6	8	48
7	3	21
$\displaystyle\sum_{i=1}^{n} f_i = 36$		$\displaystyle\sum_{i=1}^{n} f_i x_i = 139$

The arithmetic mean of this data is:

$$\bar{x} \;=\; \frac{\displaystyle\sum_{i=1}^{n} f_i x_i}{\displaystyle\sum_{i=1}^{n} f_i} \;=\; \frac{139}{36} = 3.86 \quad \text{(to 2 d.p.)}$$

For a *continuous* frequency distribution the *midpoint* of the *class* is taken as the *x* value.

Example 3:

Estimate the mean of the following data:

Mass (kg)	Number of students
60–62	4
63–65	9
66–68	20
69–71	13
72–74	4

Solution:

Mass (kg)	midpoint x_i	frequency f_i	$f_i x_i$
60–62	61	4	244
63–65	64	9	576
66–68	67	20	1340
69–71	70	13	910
72–74	73	4	292

$$\sum_{i=1}^{n} f_i = 50 \qquad \sum_{i=1}^{n} f_i x_i = 3362$$

The arithmetic mean of this data is:

$$\bar{x} = \frac{\sum_{i=1}^{n} f_i x_i}{\sum_{i=1}^{n} f_i} = \frac{3362}{50} = 67.24 \text{ kg}$$

arithmetic progression (AP): an arithmetic progression is a sequence or series of terms or numbers, where each term is obtained by adding the same constant to the previous term. An example of an arithmetic sequence and an arithmetic series are given below.

7, 11, 15, 19, 23, 27, ...

12 + 19 + 26 + 33 + 40 + ...

The first term of an AP is usually denoted by the letter *a* and the difference between the terms as *d*. This difference is called the *common difference*. The *n*th term of an AP can be found using the formula:

$$u_n = a + (n-1)d$$

The sum of the first *n* terms, S_n, of an AP can be found using the formula:

$$S_n = \frac{1}{2} n(a + l)$$

where l is the last term or

$$S_n = \frac{1}{2}n(2a + (n - 1)d)$$

Example:

For the arithmetic sequence

5, 12, 19, 26, 33, ...

find:

(a) The 20th term.
(b) The sum of the first 30 terms.

Solution:

First note that in this AP $a = 5$ and $d = 7$.

(a) Using $u_n = a + (n - 1)d$, with $n = 20$, $a = 5$ and $d = 7$ gives:

$$u_{20} = 5 + (20 - 1) \times 7$$
$$= 5 + 133$$
$$= 138$$

(b) Using $S_n = \frac{1}{2}n(2a + (n - 1)d)$,

with $n = 30$, $a = 5$ and $d = 7$ gives:

$$S_{30} = \frac{1}{2} \times 30 \times (2 \times 5 + (30 - 1) \times 7)$$

$$= \frac{1}{2} \times 30 \times (10 + 203)$$

$$= 3195$$

associative: an operation $\otimes$ is associative if $a \otimes (b \otimes c) = (a \otimes b) \otimes c$. The operations of addition and multiplication are associative because $a + (b + c) = (a + b) + c$ and $(a \times b) \times c = a \times (b \times c)$. The operations of subtraction and division are not associative.

asymptote: some curves get ever closer to a straight line, as shown in the diagram below, where the curve gets closer and closer to the broken line. This line is an asymptote to the curve.

Rational functions usually have two asymptotes, one of which is vertical and one of which is horizontal. The graph on page 14 shows

$$y = \frac{2x + 1}{x - 3}$$

This curve has two asymptotes, the lines $y = 2$ and $x = 3$.

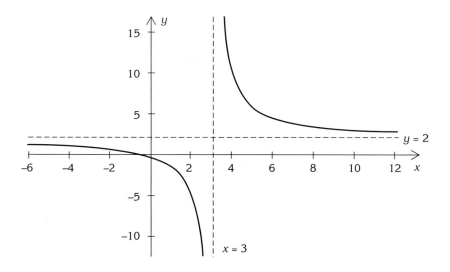

attractive force: see *central force*.

average: the name given to a number of measures (namely the *arithmetic mean*, the *geometric mean*, the *median* and the *mode*) which locate a typical value about which a distribution is clustered.

average speed: the average speed of an object is the total distance traveled by the object divided by the total time taken. See the following example.

Example:

A car travels between City A and City B, a distance of 45 miles at an average speed of 50 mph. On the return journey the average speed is 40 mph. Calculate the average speed for the total journey of City A to City B to City A.

Solution:

The total distance traveled is 90 miles. To find the average speed of the car we need to find the total time for the two journeys.

From City A to City B the average speed is 50 mph so the journey took 45/50 = 0.9 hours.

From City B to City A the average speed is 40 mph so the journey took 45/40 = 1.125 hours.

So the total journey time is 0.9 + 1.125 = 2.025 hours

The average speed is:

$$\frac{\text{total distance}}{\text{total time}} = \frac{90}{2.025} = 44.4\dot{4} \text{ mph}$$

(Note that the average speed is not the average of the two speeds.)

axis of symmetry: a shape that has rotational symmetry can be rotated about its axis of symmetry to another position in which it appears to be the same as the original position. For example a cuboid has an axis of symmetry through the center of each face as shown in the diagram below.

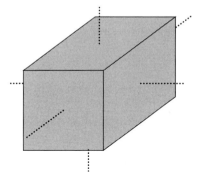

back substitution stage: a stage in the *Gaussian elimination method* in which we work backwards through the equations solving for the variables.

bar chart: a bar chart is a diagrammatic illustration of data. It is used for *categorical data*. It consists of gapped bars of equal width, their height proportional to the frequency or value they represent (see below).

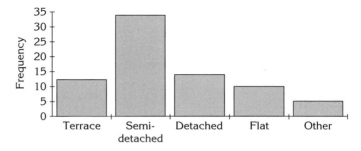

Accommodation

base of a logarithm: logarithms can be defined to any base, using the definition of a logarithm to the base a as:

$$\log_a (a^n) = n$$

Note that $\log_a (a) = \log_a (a^1) = 1$

It is most common to use base 10 or base e, and these logarithms can be found on most calculators.

It is possible to change the base of a logarithm using the result given below:

$$\log_b (x) = \frac{\log_a (x)}{\log_a (b)}$$

Examples:

(a) Find $\log_8 (x)$ given that $\log_2 (x) = 5$

(b) Show that $\log_a (b) = \dfrac{1}{\log_b (a)}$.

Solution:

(a) Using $\log_b (x) = \dfrac{\log_a (x)}{\log_a (b)}$ gives,

$$\log_8(x) = \frac{\log_2(x)}{\log_2(8)}$$

$$= \frac{5}{\log_2(2^3)} = \frac{5}{3}$$

(b) Using $\log_b(x) = \dfrac{\log_a(x)}{\log_a(b)}$ with $x = a$ gives,

$$\log_b(a) = \frac{\log_a(a)}{\log_a(b)}$$

$$= \frac{1}{\log_a(b)}$$

$$\Rightarrow \quad \log_a(b) = \frac{1}{\log_b(a)}$$

---

bearing: a bearing is the angle in a clockwise direction from North of one point from a base or origin.

For example, the bearing of Cambridge from London is 11°, the bearing of Birmingham from London is 315° and the bearing of Portsmouth from London is 225°. Here London is the origin.

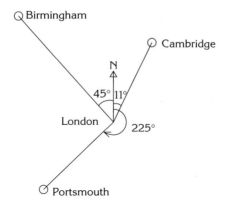

Choosing Birmingham as the origin, then the bearing of London is 135°.

A bearing is used to describe the direction of a straight line path between a starting point and a destination. For example, if I decide to walk on a bearing of 90° then I would walk due East.

bin: the "space" forming the containers for *bin-packing*.

bin-packing: a category of problems involving packing a number of objects into different containers, e.g. recording television programs of different length onto a minimum number of video tapes.

binomial distribution: a binomial situation is one where there are only two possible, *mutually exclusive*, outcomes. The outcomes are often labeled "success" or "failure." Tossing a coin is a binomial situation. There are only two possible outcomes, heads or tails.

binomial expansions

A *discrete*, random variable X with a *probability density function* (pdf):

$$P(X = x) = \binom{n}{x} p^x (1 - p)^{n-x}$$

where $x = 1, 2, 3, \ldots, n$, p is the probability of success and n is the number of trials, is said to have a binomial distribution.

$\binom{n}{x}$ is the number of *combinations* of x objects from n objects. It is also written ^nC_x.

If the distribution of X is modeled by a binomial distribution then we write $X \sim \text{Bin}(n, p)$ where n and p are known as the *parameters of the distribution*.

Example:

A row of 10 petunias is planted and the nursery gives the probability of a plant flowering to be 0.9. What is the probability that more than eight of the plants flower?

Solution:

This is a binomial situation with
$$n = 10 \text{ and } p = P(\text{success}) = P(\text{flowering}) = 0.9.$$
Let x be the number of plants flowering. Then:

$$P(X = x) = \binom{10}{x}(0.9)^x (0.1)^{10-x}$$

$$P(\text{More than 8 flowering}) = P(X = 9) + P(X = 10)$$

$$= \binom{10}{9}(0.9)^9 (0.1)^1 + \binom{10}{10}(0.9)^{10} (0.1)^0$$

$$= 0.387 + 0.349 = 0.736$$

If $X \sim \text{Bin}(n, p)$ then the *expectation* of X is $E[X] = np$ and the *variance* of X is:

$$\text{Var}[X] = np(1 - p)$$

For the binomial situation above $X \sim \text{Bin}(10, 0.9)$, $E[X] = 10 \times 0.9 = 9$ and

$$\text{Var}[X] = 10 \times 0.9 \times 0.1 = 0.9$$

binomial expansions: when the *binomial theorem* is applied to expressions that contain two terms, like $(a + b)^n$ a binomial expansion is obtained. For example:

$$(x + a)^3 = x^3 + 3x^2a + 3xa^2 + a^3$$

binomial recurrence formula: the binomial recurrence formula is given by:

$$P(X = x + 1) = \frac{n - x}{x + 1} \cdot \frac{p}{1 - p} \cdot P(X = x)$$

It is used to calculate successive probabilities once the initial probability is known.

Example:

For $X \sim \text{Bin}(10, 0.9)$, use the binomial recurrence formula to calculate $P(X = 1)$, $P(X = 2)$ and $P(X = 3)$.

Solution:

$$P(X = 0) = \binom{10}{0}(0.9)^0 (0.1)^{10} = 1 \times 10^{-10}$$

Using the recurrence formula:

$$P(X = 1) = \frac{10 - 0}{0 + 1} \times \frac{0.9}{(0.1)} \times P(X = 0) = \frac{10}{1} \times \frac{0.9}{(0.1)} \times 1 \times 10^{-10}$$

$$= 9 \times 10^{-9}$$

$$P(X = 2) = \frac{10 - 1}{1 + 1} \times \frac{0.9}{(0.1)} \times P(X = 1) = \frac{9}{2} \times \frac{0.9}{(0.1)} \times 9 \times 10^{-9}$$

$$= 3.645 \times 10^{-7}$$

$$P(X = 3) = \frac{10 - 2}{2 + 1} \times \frac{0.9}{(0.1)} \times P(X = 2) = \frac{8}{3} \times \frac{0.9}{(0.1)} \times 3.645 \times 10^{-7}$$

$$= 8.748 \times 10^{-6}$$

binomial theorem for positive integer indices: the binomial theorem states that

$$(a + b)^n = a^n + \binom{n}{1}a^{n-1}b + \binom{n}{2}a^{n-2}b^2 + \binom{n}{3}a^{n-3}b^3 + \ldots + \binom{n}{n-1}a^1 b^{n-1} + b^n$$

where n is a positive integer and $\binom{n}{r} = \dfrac{n!}{(n-r)!r!}$

For small n the values of $\binom{n}{r}$ can be found from *Pascal's triangle*.

$$
\begin{array}{ccccccccccc}
 & & & & & 1 & & & & & \\
 & & & & 1 & & 1 & & & & \\
 & & & 1 & & 2 & & 1 & & & \\
 & & 1 & & 3 & & 3 & & 1 & & \\
 & 1 & & 4 & & 6 & & 4 & & 1 & \\
1 & & 5 & & 10 & & 10 & & 5 & & 1 \\
\end{array}
$$

Example:

Expand (a) $(x + 5)^3$ (b) $\left(2 + \dfrac{x}{2}\right)^4$

Solution:

(a) Note that the coefficients needed for this expansion can be obtained from the fourth row of *Pascal's triangle*, these are 1, 3, 3, 1. These can then be used to give:

$$(x + 5)^3 = 1 \times x^3 + 3 \times x^2 \times 5 + 3 \times x \times 5^2 + 1 \times 5^3$$

$$= x^3 + 15x^2 + 75x + 125$$

(b) Note that the coefficients needed for this expansion can be obtained from the fifth row of Pascal's triangle, these are 1, 4, 6, 4, 1. These can then be used to give:

$$\left(2 + \frac{x}{2}\right)^4 = 1 \times 2^4 + 4 \times 2^3 \times \left(\frac{x}{2}\right) + 6 \times 2^2 \times \left(\frac{x}{2}\right)^2$$

$$+ 4 \times 2 \times \left(\frac{x}{2}\right)^3 + 1 \times \left(\frac{x}{2}\right)^4$$

$$= 16 + 4 \times 8 \times \frac{x}{2} + 6 \times 4 \times \frac{x^2}{4} + 4 \times 2 \times \frac{x^3}{8} + \frac{x^4}{16}$$

$$= 16 + 16x + 6x^2 + x^3 + \frac{x^4}{16}$$

binomial theorem for any rational index: the general binomial theorem can be applied to $(1 + x)^n$, for any rational n. The theorem states that:

$$(1 + x)^n = 1 + nx + \frac{n(n-1)}{2!}x^2 + \frac{n(n-1)(n-2)}{3!}x^3 + \frac{n(n-1)(n-2)(n-3)}{4!}x^4 + \ldots$$

If n is a positive integer the expansion terminates, but if n is not a positive integer it produces an infinite series. This series will converge if $-1 < x < 1$.

Example:

(a) Find the first four terms of the expansion of $(1 + x)^{-2}$
(b) Find the first four terms of the expansion of $(1 + 2x)^{1/2}$ and state the range of values of x for which it converges.

Solution:

(a) $(1 + x)^{-2} = 1 + (-2)x + \frac{(-2) \times (-3)}{2!}x^2 + \frac{(-2) \times (-3) \times (-4)}{3!}x^3 + \ldots$

$$= 1 - 2x + 3x^2 - 4x^3 + \ldots$$

(b) $(1 + 2x)^{1/2} = 1 + \frac{1}{2}(2x) + \frac{(\frac{1}{2}) \times (-\frac{1}{2})}{2!}(2x)^2 + \frac{(\frac{1}{2}) \times (-\frac{1}{2}) \times (-\frac{3}{2})}{3!}(2x)^3 + \ldots$

$$= 1 + x - \frac{1}{2}x^2 + \frac{1}{2}x^3 + \ldots$$

This series will converge if $-1 < 2x < 1$, that is if $-\frac{1}{2} < x < \frac{1}{2}$

bipartite graph: a *graph* which can be divided into two subsets in such a way that each *edge* of the graph joins a *vertex* from one subset to a vertex in the other. These graphs are particularly useful when solving problems which involve matching one set of objects to another set of objects, e.g. allocating lessons to teachers in a school timetable.

Teacher	Subjects
Miss Ardagh	Geology, Geography
Dr Berry	Geology, Mathematics, Physics
Mrs Chapman	Geography, Art
Mr Denton	Geography, Mathematics, Physics
Mrs Eeles	Art, Mathematics, Physics

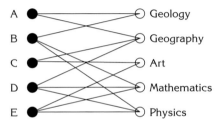

A ●   ○ Geology
B ●   ○ Geography
C ●   ○ Art
D ●   ○ Mathematics
E ●   ○ Physics

bisector: a bisector is a line that cuts either an angle or a line into two equal parts. (See also *perpendicular bisector* and *angle bisector*.)

bivariate distributions: if two variables are used to describe a population then the population is said to have a bivariate distribution. For example, if a sample of people is taken and their heights and weights recorded, the population is said to have a bivariate distribution, since it is characterized by two variables.

Boolean algebra: the algebra of symbolic logic named after the mathematician George Boole (1815–1864).

bound vector: see *free vector*.

boundary conditions: information that is given about the particular solution of a differential equation. For example "$y = 5$ when $x = 2$ and $y = 3$ when $x = 4$" are boundary conditions that could be used to find the value of the arbitrary constant in the general solution of a differential equation.

box and whisker diagram: this type of diagram is used to illustrate a *measure of location*; the *median* and *measures of spread*; the *quartiles* and *range* of a data set.

Example:

Construct a box and whisker plot of the following data: a random sample of 19 students is taken and their height in centimeters recorded:

124	142	181	192	129	141	172	157	105	135
127	196	164	163	179	148	143	154	169	

Solution:

The median of this data is 154 cm, the lower quartile, Q_1 is 135 cm, the upper quartile, Q_3 is 172 cm, the largest value is 196 cm and the smallest value is 105 cm.

The "box" is constructed from the median and quartiles while the "whiskers" are the largest and smallest values, as shown below:

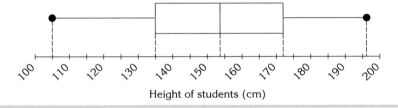

Height of students (cm)

bubble sort: an *algorithm* for sorting numbers into ascending or descending order. The idea behind the bubble sort is to place one number in its correct position in the list so that numbers that are below their correct positions rise up to their proper place (like bubbles in a fizzy drink).

bubble sort algorithm:

> **Step 1:** compare the first two numbers.
>
> **Step 2:** if the second number is larger than the first, exchange the number.
>
> **Step 3:** compare the next two numbers and repeat step 2.
>
> **Step 4:** repeat steps 1 to 3 for all pairs of numbers until you reach the end of the list.
>
> **Step 5:** repeat steps 1 to 4 until no more exchanges can be made.

The first pass using the bubble sort algorithm:

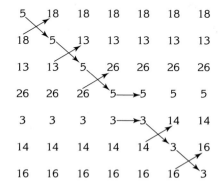

calculus: this branch of mathematics was largely due to the work of Newton and Leibnitz, and in the context of A level mathematics is mostly associated with *differentiation* and *integration*. The first of these is concerned with rates of change and particularly for finding the *gradient* or *slope* of a curve. Differentiation is used in mechanics to describe how physical quantities change. For example the velocity of an object is the rate of change of displacement.

Integration is in the main concerned with summing quantities. An application of integration is finding areas enclosed by a curve and can be applied to quantities that can be represented by these areas. Integration and differentiation are converse processes.

Cartesian coordinates: (rectangular coordinates) describe the position of any point on a plane relative to an origin and two perpendicular axes. The axes are usually referred to as the x and y axes. The position of a point (a, b) is found by starting at the origin and moving a units parallel to the x axis and then b units parallel to the y axis. The diagram shows the point (a, b), and other points which illustrate the convention used when either or both of a and b are negative.

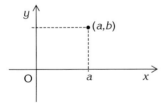
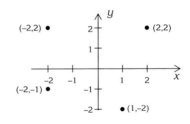

Any point in space, P, can be related to axes by three numbers a, b, c, the magnitudes of which are the perpendicular distances of P from three planes Oyz, Oxz and Oxy.

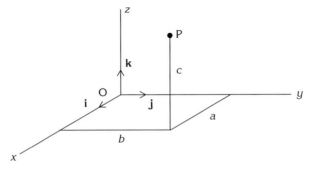

The numbers (a, b, c) are called Cartesian *coordinates* of P. This is the Cartesian coordinate system.

23

Three unit vectors **i**, **j** and **k** are usually defined along the three axes Ox, Oy and Oz. The position of P is then defined by the position vector

$$r = a\mathbf{i} + b\mathbf{j} + c\mathbf{k}$$

cascade activity number: the numbers allocated to activities in a project to ensure a logical sequence through the activities.

cascade chart: a diagram showing the precedence relations, in the form of bars drawn against a time scale, that helps to make the best use of all resources.

categorical data: data which can be classified by categories. The categories are "types" of things, for example, colors, types of accommodation, countries, etc. Categorical data are best represented pictorially in *bar charts*, *pie charts* or *pictograms*.

center of gravity: the point at which the force of gravity on a body is taken to act.

If a body is suspended from a point O by a string attached to any point P on the body, then the string will be along a vertical line passing through the center of gravity, G. The position of the center of gravity for a plane lamina can be found experimentally by drawing lines of OP produced on the lamina for different points P.

The position of the center of gravity is not necessarily at a point belonging to the body. For instance, the center of gravity, G, of a semicircular arc of radius r lies on the axis of symetry OP at a distance $2r/\pi$ from O.

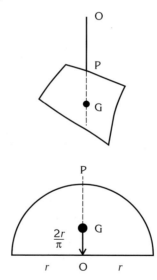

The following table shows the position of the center of gravity for various familiar objects of uniform density:

Object	Position of center of gravity
uniform semicircle	$\dfrac{4r}{3\pi}$ from the center on axis of symmetry
uniform sector of a circle, angle 2α	$\dfrac{2r\sin\alpha}{3\alpha}$ from the center on axis of symmetry

continued

Object	Position of center of gravity
uniform arc of a circle, angle 2α	$\dfrac{r \sin \alpha}{\alpha}$ from the center on axis of symmetry
uniform solid hemisphere	$\dfrac{3r}{8}$ above the center of the base
uniform solid cone	$\dfrac{h}{4}$ above the center of the base

center of mass: the point within a body at which all the mass can be considered to be.

If the force of gravity per unit mass is constant (or more generally if the gravitational field is uniform) then the positions of the center of mass and the center of gravity are identical.

For a system of n particles m_1 m_2 ... , m_n and position vectors $\mathbf{r}_1$, $\mathbf{r}_2$ ... ,$\mathbf{r}_n$ the position vector of the center of mass is

$$\mathbf{r}_G = \frac{\sum\limits_{i=1}^{n} m_i \mathbf{r}_i}{\sum\limits_{i=1}^{n} m_i}$$

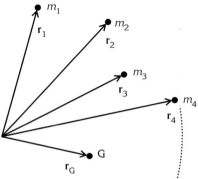

Example:

Find the center of mass of a uniform square plate ABCD of mass 4 kg and side length 0.5 m with masses of 1 kg, 2 kg, 3 kg and 4 kg at the four corners A, B, C and D, respectively.

Solution:

The diagram below shows the plate and the four masses.

Choose the corner A as the origin then the position vectors of A, B, C, D and the center O are 0, 0.5**i**, 0.5(**i** + **j**), 0.5**j** and 0.25(**i** + **j**).

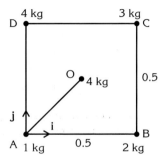

The position vector of the center of mass is

$$r_G = \frac{1 \times 0 + 2 \times (0.5i) + 3 \times 0.5(i + j) + 4 \times (0.5j) + 4 \times 0.25 (i + j)}{14}$$

$$= \frac{3.5i + 4.5j}{14}$$

$$= 0.25i + 0.32j$$

This diagram shows the position of the center of mass relative to corner A.

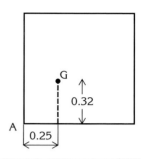

central force: a force whose direction is towards a fixed point and whose magnitude depends on the distance between the particle and that fixed point.

In the figure below, **F** is a central force acting on the particle P. The force is directed along PO and has magnitude equal to $f(r)$ where r is the distance from O to P. In vector notation the force is written as $F = -f(r)e_r$ where e_r is a unit vector along OP.

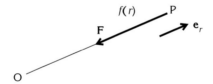

The gravitational force of the Sun on the Earth (or the Earth on the Moon) is an example of a central force. This is described by Newton's law of *gravitation*.

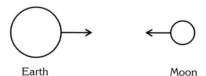

Earth   Moon

central limit theorem: let X be a random variable from any distribution with mean μ and variance σ^2. If $\overline{X}$ is the mean of a random sample of size n chosen from the distribution of X then, by the central limit theorem, $\overline{X}$ has an approximate *normal distribution* with mean μ and variance $\dfrac{\sigma^2}{n}$ for large n.

This normal approximation to any distribution has the advantage that the cumulative probabilities are easier to calculate from a normal distribution than from a *discrete* distribution such as the *binomial* or *Poisson*.

Example:

Ten dice are thrown and the number of sixes scored recorded. The ten dice are thrown 30 times. Find the probability that the average number of sixes thrown is less than two.

Solution:

This is a well-known binomial situation with the number of trials $n = 10$, the probability of success $p = \frac{1}{6}$ and the probability of failure $q = (1 - p) = \frac{5}{6}$. We can use the normal approximation to the binomial since we can regard the 30 throws as a random sample of 30. The mean and variance of a Binomial distribution are $\mu = np$ and $\sigma^2 = npq$, respectively. For this example $\mu = 1.667$ and $\sigma^2 = 1.389$.

Therefore by the central limit theorem

$$\bar{X} = \frac{\sum\limits_{i=1}^{30} X_i}{30} \sim N\left(1.667, \frac{1.389}{30}\right) = N(1.667, 0.0463)$$

The probability that the average number of sixes thrown is less than 2 is

$$P(\bar{X} < 2) = P\left(Z < \frac{2 - 1.667}{\sqrt{(0.0463)}}\right)$$

$$= P(Z < 1.549)$$

$$= 0.9393$$

For details on this probability calculation see *normal distribution*.

centrifugal force: see *inertial forces*.

centripetal force: if an object is traveling in a circle, the component of the resultant force acting on the body towards the center of the circle is called the centripetal force.

The centripetal force is necessary to maintain circular motion and the acceleration $a\omega^2$ or v^2/a towards the center, where a is the radius of the circle, v is the speed and ω is the angular speed of the object.

As an example, consider the motion of a child on a swing chair ride at a fairground. This is modeled in mechanics by a *conical pendulum* of mass m.

The actual forces acting on the child (and seat) are the tension in the chains T and the force of gravity mg. The resultant force acting on P is the horizontal component of the tension $T \sin \alpha$ along PO.

(The vertical component of the tension $T \cos \alpha$ balances the force of gravity mg.)

$T \sin \alpha$ is an example of centripetal force. Note that it acts inwards from P to O; i.e. towards the center of the circle.

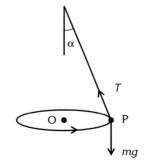

More generally, if a body moves along a curved path then the centripetal force is the component of the resultant force in the direction of the center of curvature of the curved path.

centroid: the centroid is the geometric center of a surface or body. If the surface or body is uniform (its mass is distributed evenly) then the centroid and *center of mass* coincide.

In a triangle, when lines are drawn from each corner to the midpoints of the opposite sides, the lines will intersect at the same point. This point is the centroid of the triangle.

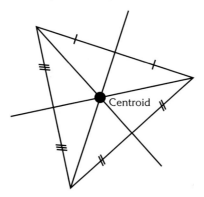

centroid of a triangle: see *centroid*.

chain rule: this is used when differentiating a function of a function, such as $y = (x^2 + 3)^6$ or $y = \sin(x^5)$. Both of these can be written in the form $y = f(u)$ where $u = g(x)$. For $y = (x^2 + 3)^6$, these functions are $f(u) = u^6$ and $g(x) = x^2 + 3$. To obtain the derivative of y the chain rule must be applied. The chain rule states:

$$\frac{dy}{dx} = \frac{dy}{du} \times \frac{du}{dx}$$

Example:

Differentiate

(a) $y = (x^2 + 3)^6$

(b) $y = \sin(x^5)$

Solution:

(a) Here $y = u^6$ and $u = x^2 + 3$. These can be differentiated to give

$$\frac{dy}{du} = 6u^5 \text{ and } \frac{du}{dx} = 2x.$$

Then the chain rule can be used to find $\frac{dy}{dx}$

$$\frac{dy}{dx} = \frac{dy}{du} \times \frac{du}{dx}$$

$$= 6u^5 \times 2x$$

$$= 12x(x^2 + 3)^5$$

(b) Here $y = \sin(u)$ and $u = x^5$. Differentiating each of these gives

$$\frac{dy}{du} = \cos(u) \text{ and } \frac{du}{dx} = 5x^4$$

Then the chain rule can be applied to give

$$\frac{dy}{dx} = \frac{dy}{du} \times \frac{du}{dx}$$

$$= \cos(u) \times 5x^4$$

$$= 5x^4 \cos(x^5)$$

chaos: some systems display extreme sensitivity to initial data and are said to exhibit "chaotic motion." An elegant view of chaos is the notion that a butterfly flapping its wings in South America could lead to a major storm in the United Kingdom. The chaotic behavior of a system restricts long range weather forecasting and the simulation of systems in biology, economics and medicine.

chi squared (χ²) statistic: used in the chi squared test the χ^2 statistic is a measure of the difference between observed (O_i) and expected (E_i) frequencies. It is calculated by the formula

$$\chi^2 = \sum_{i=1}^{n} \frac{(O_i - E_i)^2}{E_i}$$

Any expected frequencies less than five need to be combined with neighboring class frequencies.

chi squared test (χ²): the χ^2 test measures the difference between observed and expected frequencies. The expected frequencies are calculated on the basis of the *null hypothesis* H_0 that there is no difference between the two. The χ^2 statistic is computed and compared to a *critical value* of the χ^2 distribution with v *degrees of freedom* and a chosen *significance level* (normally 0.05 or 0.1).

If the calculated value of χ^2 is greater than the critical value, the observed frequencies differ significantly from the expected frequencies and H_0 is rejected in favor of an *alternative hypothesis*, H_1.

If the calculated value is less than the critical value, H_0 would be accepted (or at least not rejected). The degrees of freedom v are calculated by $v = k - m - 1$ where k is the number of categories and m is the number of parameters that need to be estimated from the observed frequencies to calculate the expected frequencies. (See also *contingency tables* and *goodness of fit*.)

Example:

A coin is tossed 100 times and the number of tails was 40. Test at the 5% level whether the coin is biased.

Solution:

H_0: the coin is unbiased.
H_1: the coin is biased.
Under the null hypothesis the expected number of heads is 50 and tails is 50.
Draw up the following table:

	O	E	$(O - E)$	$(O - E)^2$	$\dfrac{(O - E)^2}{E}$
Heads	60	50	10	100	2
Tails	40	50	−10	100	2

The χ^2 test statistic is $\chi^2 = \displaystyle\sum_{i=1}^{n} \frac{(O_i - E_i)^2}{E_i} = 4$

The degrees of freedom $v = 2 - 0 - 1 = 1$.

The 5% critical value with 1 degree of freedom is 3.84 from tables.
Since our calculated value is more than the critical value we reject H_0 and so can conclude that the coin is biased.

Chinese postman problem: see *route inspection problem*.

chord: this is a straight line that joins any two points on a curve, as shown in the diagram below.

Chord

circle: a circle is a set of points which are the same distance from a given fixed point called the center of the circle. The area of a circle of radius r is πr^2 and its circumference is $2\pi r$. The equation of a circle of radius r which has its center at the point (a, b) is:

$$(x - a)^2 + (y - b)^2 = r^2$$

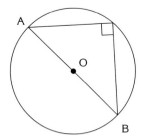

The parametric equations of the same circle are:

$$x = r\cos t + a \quad \text{and} \quad y = r\sin t + b \text{ where } 0 \le t \le 2\pi$$

circle properties: include the important results listed below.

The angle in a semicircle is always a right angle. Note that in this diagram the line AB is a diameter.

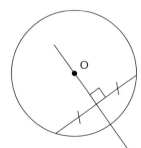

The perpendicular bisector of a chord passes through the center of the circle, marked O in the diagram.

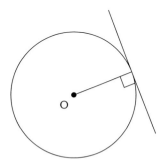

The tangent to a circle is always perpendicular to the radius that meets the circle at the same point as the tangent.

circular measure: this term describes the use of radians to measure angles. See *radian* for more detail.

circular motion: for motion in a circle, it is convenient to use polar coordinates so that $r =$ constant. The radial and transverse components of position, velocity and acceleration are shown in the diagrams on the next page:

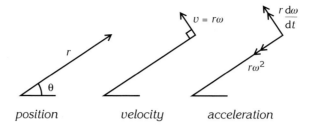

| position | velocity | acceleration |

The rate of change of θ, $\dfrac{d\theta}{dt} = \dot{\theta}$, is called the angular speed and is often denoted by ω. The velocity then has magnitude $v = r\omega$ tangential to the circle. The component of acceleration towards the center of the circle is $r\omega^2 = v^2/r$.

For *uniform circular motion* the angular speed ω is constant and the transverse component of acceleration is zero.

For uniform circular motion, the acceleration is towards the center, O.

The time taken for one complete revolution, called the period of the circular motion, is $2\pi/|\omega|$.

class: when summarizing raw data it is often useful to distribute the data into classes or categories. The number of individuals belonging to each class is known as the "class frequency." The "class interval" is the particular class and the "class limits" are the upper and lower boundaries of the class. The "class width" is the difference between the upper and lower class limit. For example, below is a frequency distribution of the mass of 50 male students recorded to the nearest kilogram.

Mass (kg)	Number of students
60–62	4
63–65	9
66–68	20
69–71	13
72–74	4

In this example the frequency in the class interval 60–62 is 4, the upper class limit is 62 and the lower class limit is 60. The class boundaries are 59.5 and 62.5, since mass is a *continuous variable*. The class width is the difference between the boundaries of the class; therefore, the width of the class interval 60–62 is 62.5 – 59.5 = 3.

cluster sampling: in cluster *sampling*, the members of the sample form a natural group or cluster. For example, when students in a college are being sampled, all the students will belong to the same cluster, which in this case would be the college.

codomain: See *function*.

coefficient of friction: two surfaces in contact experience a normal reaction force R and a tangential force F called the force of friction.

The force of friction opposes the direction of relative motion between the surfaces.

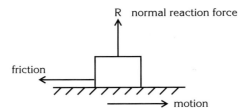

R   normal reaction force

friction

motion

For a static particle, the force of friction is just sufficient to stop motion; the law of friction is $F \leq \mu_s R$ where μ_s is the coefficient of static friction formulated by Coulomb.

For a moving particle, $F = \mu_d R$, where μ_d is the coefficient of dynamic friction.

There is experimental evidence that $\mu_d < \mu_s$.

coefficient of restitution: the constant in *Newton's law for collisions* which states that

u_1 u_2

before impact

v_1 v_2

after impact

when two bodies collide their relative velocity of separation in the direction along the normal at the point of contact is proportional to their relative velocity of approach.

If u_1 and u_2 are the velocities of the bodies before the impact; v_1 and v_2 are the velocities of the bodies after the impact, then

$$v_2 - v_1 = -e(u_2 - u_1)$$

The constant e is called the coefficient of restitution and $0 \leq e \leq 1$. If $e = 0$ the collision is inelastic and the bodies adhere after impact; if $e = 1$ the collision is perfectly elastic.

Example:

Two spheres of masses 0.4 kg and 0.6 kg moving in the same direction with speeds 3 m s^{-1} and 2 m s^{-1} collide. Find their subsequent speeds after the impact if the coefficient of restitution is 0.8.

Solution:

0.4   0.6

3 ms^{-1} 2 ms^{-1}

before impact

0.4   0.6

v_1 v_2

after impact

33

To solve problems of this type we apply two laws:

conservation of momentum:

$$0.4v_1 + 0.6v_2 = 0.4 \times 3 + 0.6 \times 2 = 2.4$$

Newton's law for collisions:

$$v_2 - v_1 = -e\,(u_2 - u_1) = -0.8(2 - 3) = 0.8$$

Solving these two equations for v_1 and v_2 gives

$$v_1 = 1.92 \text{ and } v_2 = 2.72$$

collinear: a set of points is described as collinear if the points all lie on the same straight line. For example the points (1, 4), (2, 7) and (3, 10) are collinear because they are all on the straight line with the equation $y = 3x + 1$, as shown in the diagram below.

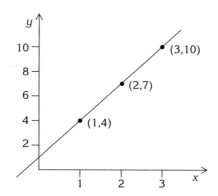

collisions: see *conservation of momentum, coefficient of restitution* and *impulse.*

combinations: the number of different combinations or selections of r objects which can be made from n distinct objects is given by the formula:

$$^nC_r = \binom{n}{r} = \frac{n!}{r!(n-r)!}$$

Example:

A group of six people wishes to choose two from their number to run an errand. How many ways can the two be chosen from the six?

Solution:

There are several different ways to choose two people and this is called the number of combinations. If the six people are A, B, C, D, E, and F then the possible selections are:

A and B	A and C	A and D
A and E	A and F	B and C
B and D	B and E	B and F
C and D	C and E	C and F
D and E	D and F	E and F

There are 15 possible combinations of the two people.

This can be confirmed by using the formula

$$^6C_2 = \frac{n!}{r!(n-r)!} = \frac{6!}{2!(6-2)!} = \frac{6!}{2!4!} = \frac{720}{2 \times 24} = 15$$

combined events: in probability theory, if two or more *events* occur together they are called combined events. A *tree diagram* is a good way of finding the probability of a combined event.

As an example, consider the two experiments "tossing a coin" and "throwing a die" and the two independent events: "getting a head, H" and "getting a five, F."

The two events are combined events and are written as H ∩ F

The probability of getting a head P(H) = ½

The probability of getting a five P(F) = ⅙

Therefore P(H ∩ F) = ½ × ⅙ = ¹⁄₁₂

common chord: when two circles intersect at two points, say P and Q, the line PQ is called the common chord. This is shown in the diagram below.

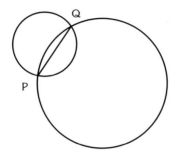

common difference: describes the difference between successive terms of an *arithmetic progression*. This is usually denoted by the letter d. The common difference for the sequence 7, 11, 15, 19, 23, ... is 4 and the common difference for the sequence 90, 85, 80, 75, 70, ... is –5.

common factor: a common factor is a number, a polynomial or a quantity that is a factor of each member of a given set of objects.

Example:

Find a common factor of the following:

(a) {49, 147, 350}
(b) {$(x^2 + 3x + 2)$, $(x^2 - 1)$}

Solution:

For each set we break the members down into a product of factors.

(a) 49 = 7^2
    147 = 3×7^2
    350 = $2 \times 5^2 \times 7$

A common factor of the set {49, 147, 350} is 7.

(b) $(x^2 + 3x + 2) = (x + 2)(x + 1)$
$\quad\ \ (x^2 - 1) = (x - 1)(x + 1)$

A common factor of the set $\{(x^2 + 3x + 2), (x^2 - 1)\}$ is $(x + 1)$

common logarithms: *logarithms* to the base 10. They have the property that $\log_{10}(10^n) = n$. For example:

$$\log_{10}(100000) = \log_{10}(10^5) = 5$$

and

$$\log_{10}\left(\frac{1}{100}\right) = \log_{10}(10^{-2}) = -2$$

common ratio: in a *geometric progression* the common ratio is the ratio between successive terms of the progression and is usually denoted by r. For example in the sequence, 9, 3, 1, $\frac{1}{3}$, $\frac{1}{9}$, $\frac{1}{27}$, ... the common ratio is $\frac{1}{3}$ and in the sequence 2, –4, 8, –16, ... the common ratio is –2.

commutative: an operation $\otimes$ is commutative on some set S if $a \otimes b = b \otimes a$ for all a, b in S. The operations of addition and multiplication are commutative because $a + b = b + a$ and $a \times b = b \times a$. The operations of division and subtraction are not commutative because $a - b \neq b - a$ and $a \div b \neq b \div a$.

complement of a set: the complement of a set E, written $\bar{E}$, is the subset of the outcome set where E does not occur. The *Venn diagram* below shows outcome set S and its subsets E and $\bar{E}$.

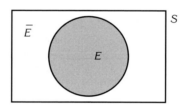

For example, a die is thrown once. The outcome set S is the set of all possible outcomes: $\{1, 2, 3, 4, 5, 6\}$. Let E be the event "multiples of 2," then $E = \{2, 4, 6\}$. Therefore the complement of E is $\bar{E} = \{1, 3, 5\}$.

complete bipartite graph: a *bipartite graph* in which each *vertex* in one subset is joined to every vertex in the other by an *edge*. A complete bipartitie graph with n vertices in one set and m vertices in the other set is denoted by $K_{n,m}$.

complete graph: a *graph* where each *vertex* is connected to every other vertex. A complete graph with n vertices is denoted by K_n. For example, the complete graph K_5 has five vertices and ten *edges*:

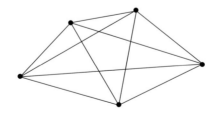

completing the square: this is a process that converts a quadratic expression of the form, $x^2 + bx + c$ into the form

$$\left(x + \frac{b}{2}\right)^2 + c - \left(\frac{b}{2}\right)^2$$

This process then makes it very easy to find the solutions of a *quadratic equation*.

It can be applied to the general form of a quadratic equation to derive the *quadratic equation formula*.

Example:

Solve the quadratic equation $x^2 + 6x + 1 = 0$, by completing the square.

Solution:

Completing the square gives:

$$x^2 + 6x + 1 = \left(x + \frac{6}{2}\right)^2 + 1 - \left(\frac{6}{2}\right)^2$$

$$= (x + 3)^2 - 8$$

Now solving the equation gives:

$$(x + 3)^2 - 8 = 0$$
$$(x + 3)^2 = 8$$
$$x + 3 = \pm\sqrt{8}$$
$$x = -3 \pm \sqrt{8}$$

complex numbers: numbers of the form $a + bi$, where a and b are real numbers and $i = \sqrt{(-1)}$. The complex number $a + bi$ has two parts, a which is called the "real part" and b which is called the "imaginary part." The complex number system allows solutions to be found for equations that do not have real solutions. For example the quadratic equation $x^2 + 4x + 5 = 0$ has solutions that are given by:

$$x = \frac{-4 \pm \sqrt{(4^2 - 4 \times 1 \times 5)}}{2 \times 1}$$

$$= \frac{-4 \pm \sqrt{(-4)}}{2}$$

Having reached this stage it can be concluded that the equation has no real solutions. However using $i = \sqrt{(-1)}$, two complex solutions can be obtained.

$$x = \frac{-4 \pm \sqrt{(-1)} \times \sqrt{4}}{2}$$

$$= \frac{-4 \pm 2i}{2}$$

$$= -2 \pm i$$

(See also *argand diagram, addition and subtraction of complex numbers, division of complex numbers* and *multiplication of complex numbers*.)

components of a force: a component of a force is a projection of the force in a specified direction.

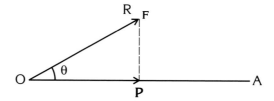

P is the component of the force **F** along the line OA. The magnitude of **P** is $F\cos\theta$ where F is the magnitude of **F**. An easy way to think of the component of a force in a given direction is:

"The magnitude of the force times the cosine of the angle you turn through"

In the figure the angle θ is the angle the force is turned through to align OR with OA.

Example:

Find the components of the forces in the following diagram in the direction of the lines OA and OB.

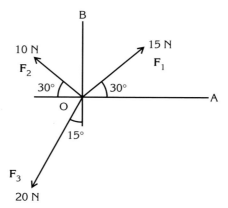

Solution:

The table shows the components

Component in the direction of

Force	OA	OB
F_1	$15\cos30° = 12.99$ N	$15\cos60° = 7.5$ N
F_2	$10\cos150° = -10\cos30° = -8.66$ N	$10\cos60° = 5$ N
F_3	$20\cos105° = -20\cos75° = -5.18$ N	$20\cos165° = -20\cos15° = -19.32$ N

It is usually convenient to resolve forces into two components with specific directions perpendicular to each other. In cartesian coordinates the specified directions are parallel to the x and y axes:

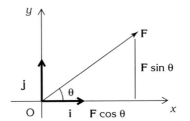

$F \cos \theta$ is the component of **F** in the x direction.

$F \cos (90 - \theta) = F \sin \theta$ is the component of **F** in the y direction.

If we choose unit vectors **i** and **j** along Ox and Oy respectively then

$$\mathbf{F} = F\cos \theta\, \mathbf{i} + F\sin \theta\, \mathbf{j}$$

or $\qquad \mathbf{F} = \begin{pmatrix} F\cos \theta \\ F\sin \theta \end{pmatrix}$ as a column vector

For the forces in the diagram with OA as the x axis we have

Force	Cartesian components of the force
F1	$12.99\,\mathbf{i} + 7.5\,\mathbf{j}$ or $\begin{pmatrix} 12.99 \\ 7.5 \end{pmatrix}$
F2	$-8.66\,\mathbf{i} + 5\,\mathbf{j}$ or $\begin{pmatrix} -8.66 \\ 5 \end{pmatrix}$
F3	$-5.18\,\mathbf{i} - 19.32\,\mathbf{j}$ or $\begin{pmatrix} -5.18 \\ -19.32 \end{pmatrix}$

components of a vector: this is the projection of a vector in a specified direction. If a vector **a** is written as the sum of two vectors **p** and **q** then:

$$\mathbf{a} = \mathbf{p} + \mathbf{q}$$

and **p** and **q** are called components of **a**.

It is usually convenient to resolve a vector into components that are parallel to the x and y axes along which unit vectors **i** and **j** are defined

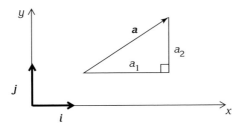

$$\mathbf{a} = a_1\mathbf{i} + a_2\mathbf{j}$$

$a_1 = a \cos \theta$ and $a_2 = a \sin \theta$ are called the Cartesian components of **a**.

The components of any vector are found in the same way as finding the *components of a force*.

composite functions: when two or more functions are combined by taking a function of a function, then the result is called a composite function. If the functions f and g are combined in this way, then the composite function is fg. Note that normally this will **not** be the same as gf.

Example:

If $f(x) = x^2 + 1$ and $g(x) = 3x - 2$,
find (a) $fg(2)$, (b) $gf(2)$, (c) $fg(x)$, (d) $gf(x)$.

Solution:

(a) $g(2) = 3 \times 2 - 2 = 4$, so $fg(2) = f(4) = 4^2 + 1 = 17$.
(b) $f(2) = 2^2 + 1 = 5$, so $gf(2) = g(5) = 3 \times 5 - 2 = 13$.
(c) $fg(x) = f(3x - 2) = (3x - 2)^2 + 1 = 9x^2 - 12x + 5$.
(d) $gf(x) = g(x^2 + 1) = 3(x^2 + 1) - 2 = 3x^2 + 1$

compound angle formulas: see *trigonometric sum and difference formulas.*

concurrent: lines that all meet at the same point are concurrent. The diagram shows two sets of lines; the lines on the left are concurrent, while the ones on the right are not.

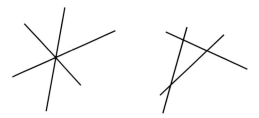

conditional probability: the conditional probability of A given B is the probability of event A occurring given that event B has already occurred. It is written P(A|B) and can be calculated from the formula:

$$P(A|B) = \frac{P(A \cap B)}{P(B)}$$

Example:

A card is pulled from a pack of 52 cards. Find the probability that the card is a heart given that it is a 5.

Solution:

Let H be the event "the card is a heart"
and 5 be the event "the card is a 5"

$$P(H|5) = \frac{P(H \cap 5)}{P(5)} = \frac{1/52}{1/13} = \frac{1}{4}$$

cone: a right circular cone has a circular base and its curved surface rises to a point directly above the center of the circular base. If a cone has the dimensions shown in the diagram on page 41, then its volume is given by $\frac{1}{3}\pi r^2 h$ and the area of the curved surface is $\pi r l$.

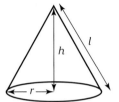

confidence intervals: given a *random sample* from a *population*, it is possible to find an interval in which an unknown parameter from the population is expected to lie with a given degree of confidence. This interval is called the confidence interval and it is defined by the "confidence limits."

If $\bar{x}$ is the mean of a random sample of size n from a population which is distributed N (μ, σ^2), and σ^2 is known, then a $B\%$ confidence interval for the population mean, μ, is given by:

$$\bar{x} - z\,\frac{\sigma}{\sqrt{n}} \;<\; \mu \;<\; \bar{x} + z\,\frac{\sigma}{\sqrt{n}}$$

where z is the $\frac{1}{2}(100 - B)\%$ point of a standard *normal distribution*.

If a large sample ($n \geq 30$), is taken from any distribution, then by the *central limit theorem*, that sample is distributed $N(\mu, (\sigma^2/n))$ where μ is the unknown population mean and σ^2 is the unknown population variance. A $B\%$ confidence interval for μ is given by

$$\bar{x} - z\,\frac{s}{\sqrt{n}} \;<\; \mu \;<\; \bar{x} + z\,\frac{s}{\sqrt{n}}$$

where $\bar{x}$ and s^2 are the sample mean and variance.

If r is the proportion of a random sample of size n from a population, a $B\%$ confidence interval for the population proportion is given by:

$$r - z\sqrt{\left(\frac{r(1-r)}{n}\right)} \;<\; \Pi \;<\; r + z\sqrt{\left(\frac{r(1-r)}{n}\right)}$$

where z is the $\frac{1}{2}(100 - B)\%$ point of a standard normal distribution.

Example:

A random sample of ten students from an eleventh-grade class was taken and their height measured. The mean height of these students was found to be 1.75 m. It is assumed that the height of the population is normally distributed with known variance of 0.06 m. Find a 95% confidence interval for the mean height of all students in the eleventh grade.

Solution:

A 95% confidence interval for μ is given by:

$$\bar{x} - z\,\frac{\sigma}{\sqrt{n}} \;<\; \mu \;<\; \bar{x} + z\,\frac{\sigma}{\sqrt{n}}$$

$$1.75 - 1.96\,\frac{0.245}{\sqrt{10}} \;<\; \mu \;<\; 1.75 + 1.96\,\frac{0.245}{\sqrt{10}}$$

$$1.75 - 0.15 \; < \; \mu \; < \; 1.75 + 0.15$$

$$1.6 \; < \; \mu \; < \; 1.9$$

The 95% confidence interval is $1.6 < \mu < 1.9$ or $\mu \pm 0.15$. The confidence limits are 1.6 and 1.9. We can say that the probability that the interval contains the population mean is 0.95.

confidence limits: see *confidence interval*.

congruent: two shapes are described as congruent if they are identical. This means for example that two congruent *polygons* will have corresponding sides of the same lengths and the angles between them will be identical. Two congruent shapes could be placed so that one was on top of the other and covered it exactly.

conic: a conic section or conic is the shape obtained by the intersection of a plane and a circular cone.

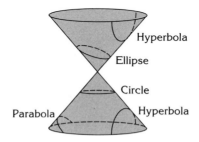

The parabola, hyperbola, ellipse and circle are curves that form the conics. The equations of all the conics can be written in Cartesian coordinates in the form

$$ax^2 + bxy + cy^2 + dx + ey + f = 0$$

where a, b, c, d, e and f are constants; and in polar coordinates the equation will have the form:

$$\frac{l}{r} = 1 + e \cos \theta$$

where l and e are constants called the semi-latus rectum and eccentricity respectively

The following table shows the standard form of the equation for each conic centered on the origin.

Conic	Cartesian equation	Polar equation
circle	$x^2 + y^2 = a^2$	$r = a$
ellipse	$\dfrac{x^2}{a^2} + \dfrac{y^2}{b^2} = 1$	$\dfrac{l}{r} = 1 + e \cos \theta, \; e < 1$
hyperbola	$\dfrac{x^2}{a^2} - \dfrac{y^2}{b^2} = 1$	$\dfrac{l}{r} = 1 + e \cos \theta, \; e > 1$
parabola	$y^2 = 4ax$	$\dfrac{l}{r} = 1 + \cos \theta, \; e = 1$

The conics are important curves in mechanics because an object in motion experiencing a gravitational central force will move along one of the conics. For example, the planets in the solar system move around the Sun along elliptic paths.

conical pendulum: a conical pendulum consists of an object suspended by a "string" from a fixed point and forced to travel in a horizontal circle.

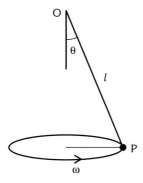

conical pendulum

The length of the string (OP) l, the angular speed of the object ω and the angle of the string to the vertical θ are related by the formula

$$\cos \theta = \frac{g}{l\omega^2}$$

where g is the acceleration due to gravity. The motion of the object P is independent of its mass.

conjugate: the conjugate of the complex number $a + bi$ is $a - bi$. For a complex number z it is usual to write z^* for its conjugate. For example if $z = 3 + 4i$, then $z^* = 3 - 4i$ and if $z = 5 - 2i$, then $z^* = 5 + 2i$. The conjugate is particularly useful in the division of complex numbers because $z \times z^*$ is always a real number. If $z = a + bi$, then $z^* = a - bi$ so that:

$$\begin{aligned} z \times z^* &= (a + bi)(a - bi) \\ &= a^2 - abi + abi - (bi)^2 \\ &= a^2 - b^2 \times (-1) \\ &= a^2 + b^2 \end{aligned}$$

(See also *division of complex numbers*.)

connected graph: a *graph* with at least one route between any pair of *vertices*.

connected particles: a term that refers to two particles that are connected by a light, inextensible string. Problems normally involve the case when the string is taut and passes over a smooth pulley or peg. In solving these problems it is important to note that the magnitude of the tension exerted by the string on each particle will be the same and that the magnitude of the acceleration of each particle will be the same. Newton's second law can be applied to each particle and the resulting equations solved to find the required quantities, for example the tension in the string and the acceleration of the particles.

Example:

The diagram shows a particle, of mass 7 kg, sliding on a rough horizontal surface and connected by a light, inextensible string to another particle, of mass 5 kg.

The coefficient of friction between the surface and the particle is 0.2. Find the acceleration of each particle and the tension in the rope.

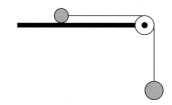

Solution:

The diagrams show the forces acting on each particle

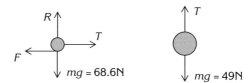

First consider the particle on the surface. From the vertical components $R = 68.6$ N and as the particle is sliding $F = 0.2 \times 68.6 = 13.72$ N.

So the resultant force, to the right, on this particle is:

$$T - F = T - 13.72$$

Using Newton's second law, $F = ma$, gives:

$$7a = T - 13.72 \qquad (1)$$

The resultant downward force on the other particle is:

$$49 - T$$

Using Newton's second law, $F = ma$, gives:

$$5a = 49 - T \qquad (2)$$

From equation (1), $T = 7a + 13.72$.

Substituting this into equation (2) gives:

$$5a = 49 - (7a + 13.72)$$
$$12a = 49 - 13.72$$
$$a = \frac{35.28}{12}$$
$$= 2.94 \text{ m s}^{-2}$$

This can now be used to find T:

$$T = 7 \times 2.94 + 13.72$$
$$= 34.3 \text{ N}$$

conservation of energy: this fundamental physical principle states that the total amount of energy of the universe is constant. Energy may be transformed from one form to another but it cannot be created or destroyed.

On a more "local level" we say that the total amount of energy of an isolated system is constant. By an isolated system we mean that no energy is allowed to escape.

In mechanics, we often refer to the "law of conservation of mechanical energy." For this law we are saying that:

kinetic energy + potential energy = constant

This law can be applied in a problem provided that such mechanical energy cannot be transferred into other forms of energy such as heat, light and sound. It is often a useful law for problems in which friction forces can be neglected.

(See also *mechanical energy, kinetic energy, potential energy*.)

conservation of momentum: this basic physical principle states that the total momentum of a system remains constant provided no external force acts on the system.

This is a remarkable law that applies to atomic particle physics as well as to vast galaxies. It says that if one part of a system loses momentum then some other part will gain it. For example, when a tennis racket hits a ball the ball gains the same amount of momentum as the racket loses; the total momentum of the racket and ball just before the impact equals the total momentum just after the impact.

Example:

A truck of mass 10 tonnes traveling with speed 15 m s^{-1} (approximately 30 mph) collides with a car of mass 1.5 tonnes traveling with speed 7.5 m s^{-1}. The truck and car are traveling along a straight road in the same direction. Find the speed of the truck and car after the collision if they become locked together.

Solution:

Before the collision

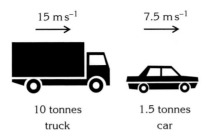

15 m s^{-1} 7.5 m s^{-1}

10 tonnes   1.5 tonnes
truck   car

total momentum = $15 \times 10\,000 + 7.5 \times 1500$

= 161 250 N s

After the collision

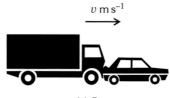

$v \, \text{m s}^{-1}$

11.5 tonnes

total momentum $= 11\,500v \, \text{N s}$

The principle of conservation of momentum gives

$$11\,500v = 161\,250$$

$$v = 14 \text{ m s}^{-1} \text{ (to 2 s.f.)}$$

The truck and car travel at a speed of 14 m s⁻¹ after the collision.

(See also *impulse* and *coefficient of restitution*.)

conservative force: a force on an object for which the work done by the force as the object moves between any two points A and B, depends only on the positions of A and B and not on the path between A and B.

An alternative definition of a conservative force is a force for which no net work is done as the particle moves along every closed path from a point A to a point B and back to A.

The force of gravity is an example of a conservative force because the work done in raising a particle through a height *h* is *mgh*. Its value does not depend on how the particle is raised through the height.

Friction is an example of a nonconservative force.

Suppose that an object slides along two paths between A and B on a horizontal surface:

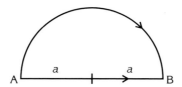

path 1 is a straight line of length 2*a*

path 2 is a semicircle with AB as diameter and radius *a*.

The friction force on the object is µ*mg*.

The work done against friction for path 1 is 2µ*mga* and for path 2 is πµ*mga* (in each case the work done is force × distance). Hence work done depends on the path taken so that friction is not a conservative force.

conservative system: a mechanical system for which the sum of the kinetic energy and potential energy is constant. (See also *conservation of energy*.)

constant acceleration equations: for motion in a straight line for which the acceleration is constant the displacement *s*, time *t*, and velocity *v* are related through the equations:

$$v = u + at$$
$$v^2 = u^2 + 2as$$
$$s = ut + \tfrac{1}{2}at^2$$
$$s = \tfrac{1}{2}(u + v)t$$

where u is the initial velocity (the value of v when $t = 0$).

These equations can be written in vector form:

$$\mathbf{v} = \mathbf{u} + \mathbf{a}t$$
$$\mathbf{v}.\mathbf{v} = \mathbf{u}.\mathbf{u} + 2\mathbf{a}.\mathbf{s}$$
$$\mathbf{s} = \mathbf{u}t + \tfrac{1}{2}\mathbf{a}t^2$$
$$\mathbf{s} = \tfrac{1}{2}(\mathbf{u} + \mathbf{v})t$$

provided the acceleration vector **a** is constant in both magnitude and direction.

constant of integration: this is a constant added to every indefinite integral that does not have limits of integration, for which the letter c is usually used. For example consider the integral of x^2:

$$\int x^2 \, dx = \frac{x^3}{3} + c$$

where c is a constant called the constant of integration.

constraints: the linear equalities or linear inequalities that restrict the values of the variables in *linear programming problems*.

contact forces (surface forces): forces are often classified into two types:

- contact forces, which occur when two surfaces are in contact
- body forces, which act at every point throughout the volume of a body.

The normal reaction and friction are examples of contact forces; the force of gravity is an example of a body force.

contingency tables: a way of summarizing the relationship between two variables. An r times c contingency table has r rows and c columns.

For example, 400 people were classified by hair color and eye color. The information is summarized in the contingency table below. Here $r = 4$ and $c = 3$.

Hair color	Eye color			
	Brown	**Grey/Green**	**Blue**	**Totals**
Black	50	54	41	145
Brown	38	46	48	132
Fair	22	30	31	83
Ginger	10	10	20	40
Totals	120	140	140	400

Often the chi squared test is used to test for independence between the two variables. The expected frequencies are calculated on the basis of the *null hypothesis* H_0, that the two variables are independent. On this basis the expected value for each cell would be:

$$\frac{\text{Row total} \times \text{Column total}}{\text{Grand total}}$$

The degrees of freedom for a r times c contingency table is $(r-1)(c-1)$.

Example:

For the contingency table above test the independence of hair color and eye color.

Solution:

H_0: eye and hair color are independent

H_1: eye and hair color are not independent

Under H_0 the expected frequencies are calculated.

Hair color	Eye color			Totals
	Brown	**Grey/Green**	**Blue**	
Black	$\frac{145 \times 120}{400} = 43.5$	$\frac{145 \times 140}{400} = 50.75$	$\frac{145 \times 140}{400} = 50.75$	145
Brown	$\frac{132 \times 120}{400} = 39.6$	$\frac{132 \times 140}{400} = 46.2$	46.2	132
Fair	24.9	29.05	29.05	83
Ginger	12	14	14	40
Totals	120	140	140	400

The χ^2 test can now be performed as in the standard case. The table below shows the values of $(O-E)^2/E$.

Hair color	Eye color		
	Brown	**Grey/Green**	**Blue**
Black	0.971	0.208	1.873
Brown	0.065	0.001	0.070
Fair	0.338	0.031	0.131
Ginger	0.333	1.143	2.571

The degrees of freedom are $(r-1)(c-1) = (4-1)(3-1) = 3 \times 2 = 6$.

The χ^2 test statistic is $\chi^2 = \sum \frac{(O-E)^2}{E} = 7.735$.

The 5% critical value with six degrees of freedom is 12.592 from tables.

Since our calculated value is less than the critical value we accept H_0 and so can conclude that hair color and eye color are independent for this sample.

When this test is performed on a graphic calculator, the p *value* obtained for $\chi^2 = 7.735$ is 0.2381, which is greater than 0.05 (the level of significance), which leads to the conclusion of accepting H_0.

continuous data: data that have no precise fixed value and are usually measured to within a range. Such data includes height (measurable only to the nearest unit, a millimeter say), or time (measurable to the nearest second for instance). (See also *discrete data*.)

continuous probability distributions: this is the probability distribution of a continuous random variable, X, which can take only values in the ranges x_1 up to x_2, x_2 up to x_3 ..., x_n up to x_{n+1} with probabilities $p_1, p_2, ..., p_n$ respectively. (See also *continuous random variable*, *probability density function*.)

continuous random variable: let X be a continuous variable which takes values in the ranges $x_1 \leq X < x_2, x_2 \leq X < x_3, ... x_n \leq X < x_{n+1}$ with probabilities $p_1, p_2 ... p_n$ respectively.

$$\text{if } \sum_{i=1}^{n} p_i = 1 \text{ then } X \text{ is a continuous random variable}$$

A continuous random variable is usually defined by a *probability density function* (*pdf*).

Example:

A continuous random variable X takes values between 0 and 5 with the following probabilities:

X	$0 \leq X < 1$	$1 \leq X < 2$	$2 \leq X < 3$	$3 \leq X < 4$	$4 \leq X < 5$
P(X)	0.1	0.2	0.4	0.2	0.1

Show that X is a continuous random variable.

Solution:

$$\sum_{i=1}^{n} p_i = 0.1 + 0.2 + 0.4 + 0.2 + 0.1 = 1$$

and therefore X is a continuous random variable.

continuous variable: see *continuous random variable*.

converge: a sequence of numbers is said to converge if the numbers get closer and closer to a specific value, known as the limit. The sequence of numbers below converges to 1, because the fractions get closer and closer to 1 as the sequence continues.

$$\frac{8}{9}, \frac{9}{10}, \frac{10}{11}, \frac{11}{12}, \frac{12}{13}, \frac{13}{14}, \frac{14}{15}, ...$$

The sequence 1, 1.1, 1.2, 1.3, 1.4, 1.5, ... does not converge as the terms continue to get larger. This sequence is said to *diverge*.

coordinates: a coordinate is a set of numbers that describes the position of a point. There are several different systems that can be used. The most common system is *Cartesian* or *rectangular coordinates* but others such as *polar coordinates* are also used.

coplanar: if a set of lines all lie in the same plane then they are coplanar. If you drew lines on the floor, they would be coplanar, as they would all lie in the plane of the floor.

correlation: a measure of how well a linear equation describes or explains the relationship between two random variables X and Y say. If Y increases linearly as X increases, we say that there is "positive linear correlation" between X and Y; if Y decreases as X increases we say that there is negative linear correlation between X and Y; if there is no relationship between X and Y then we say that there is no linear correlation between the two variables. If all the values of the variables perfectly satisfy a straight line equation we say there is "perfect correlation" between the two variables.

The amount of correlation between two variables can be measured quantitatively by calculating a correlation coefficient. There are two types of correlation coefficient – one concerned with the magnitude of the observations, for example *Pearson's correlation coefficient* (the *product–moment correlation coefficient*) and the other with their ranks, for example *Spearman's ranked correlation coefficient* or *Kendal's ranked correlation coefficient*.

The value of the correlation coefficient r takes values only in the range $-1 \leq r \leq 1$. Negative values represent negative correlation and positive values represent positive correlation. A value of r close to ± 1 indicates strong or high correlation, values closer to 0 indicate no correlation.

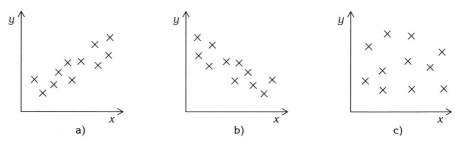

Scatter diagrams showing:

a) *positive correlation*
b) *negative correlation*
c) *no correlation*

cosecant: defined as the reciprocal of sine. This definition gives,

$$\text{cosec } \theta = \frac{1}{\sin \theta}$$

Note the cosecant of θ is abbreviated to cosec θ or csc θ. The graph below shows sin x and cosec x.

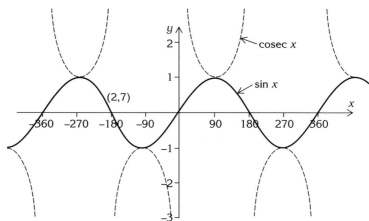

cosine:

Consider a triangle ABC in which the angle at B is a right angle. If we label the side adjacent to angle A as a and the hypotenuse as h, then we define the cosine of angle A as

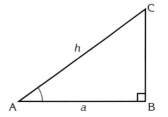

$$\cos(A) = \frac{a}{h}$$

The cosine of an angle bigger than 90° is also defined. The graph below shows the cosine function for angles between 0 and 720°.

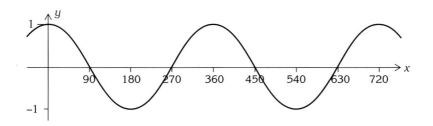

Note that the cosine function has a period of 360° (or 2π radians) and an amplitude of 1.

cosine rule: the cosine rule can be applied to any triangle to find an unknown angle or length. To find a length with the cosine rule you need to know the length of two sides of the triangle and the size of the angle between them. To find an angle the lengths of all the three sides must be known. The cosine rule can be expressed in two forms. The first used for finding lengths is:

$$a^2 = b^2 + c^2 - 2bc \cos A$$

and the second used for finding angles is:

$$\cos A = \frac{b^2 + c^2 - a^2}{2bc}$$

It is important to note the convention for labeling the sides and angles as shown in the diagram. The cosine rule can be used in conjunction with the sine rule to solve problems that are not straightforward.

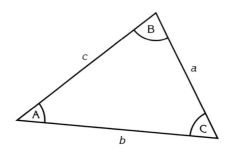

cos⁻¹: this is the inverse of the cosine function. It is used when finding an angle given its cosine, and is often used in the context of right-angled triangles. When using a calculator you will be given the principal value, which will be between 0 and 180 in degrees or 0 and π in radians. The graphs below show $y = \cos^{-1} x$ working in degrees on the left and radians on the right. Note that the function is defined only for $-1 \leq x \leq 1$. The inverse cosine function is often denoted by *arccos*.

The derivative of cos⁻¹ can be obtained by using the chain rule and a trigonometric identity. Start with $y = \cos^{-1} x$ and rearrange as $x = \cos y$. Then, differentiating with respect to x and using the chain rule gives

$$1 = -\sin y \frac{dy}{dx}, \quad \text{or} \quad \frac{-1}{\sin y} = \frac{dy}{dx}$$

It is possible to eliminate $\sin y$ from this expression by using the identity $\sin^2 y + \cos^2 y = 1$. As $\cos y = x$ this gives

$$\sin y = \sqrt{(1 - \cos^2 y)} = \sqrt{(1 - x^2)}.$$

Replacing $\sin y$ in the expression for $\frac{dy}{dx}$ gives $\frac{dy}{dx} = \frac{-1}{\sqrt{(1 - x^2)}}$, as the derivative of cos⁻¹ x.

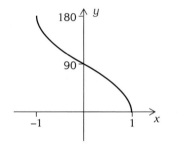 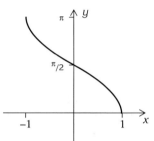

cosec⁻¹: this is the inverse of *cosec*.

cost function: a relationship linking the cost to the variables that are included in the cost.

cost matrix: a matrix representing the *weights* on the *edges* of a *network*.

	A	B	C	D	E
A	–	8	5	10	–
B	8	–	–	3	5
C	5	–	–	4	7
D	10	3	4	–	4
E	–	5	7	4	–

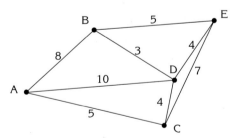

cost paths: *shortest path problems* involve finding paths with optimal properties, not necessarily to do with lengths. If the optimal property is the cheapest way it is called a cost path.

cotan⁻¹: this is the inverse of the *cotangent (cot)*.

cotangent (cot): this is the reciprocal of tangent. The cotangent of angle θ is usually written as cot θ and defined as:

$$\cot \theta = \frac{1}{\tan \theta} = \frac{\cos \theta}{\sin \theta}$$

couple: a pair of forces which are equal in magnitude, opposite in direction and do not act through the same point.

For two forces of magnitude *F* whose lines of action are distance *d* apart, then the moment of the couple about any point has magnitude *Fd*.

Two couples are said to be equivalent if their moments are the same.

(See also *moment of a force*.)

critical activities: those activities that must be started on time to avoid delaying the whole project in a *critical path* analysis.

critical path: the minimum time needed to complete a project, following the longest path through the *precedence network* and including all the *critical activities*.

critical region: in hypothesis testing, if the *test statistic* lies in the critical region (which is determined by the type of test and level of *significance*) then the *null hypothesis* is rejected. This region can always be shown in a diagram. The "critical value" is the boundary of the critical region.

For a two-tailed test of a *normal distribution*:

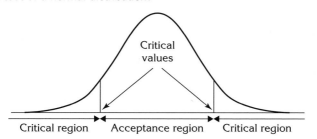

and for one-tailed tests:

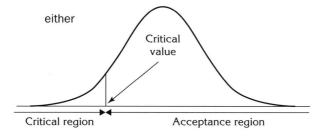

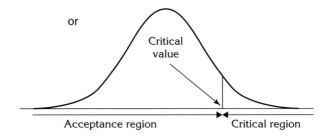

or

Critical value

Acceptance region   Critical region

Example:

A coin is tossed 100 times and the number of tails is 40. Test at the 5% level whether the coin is biased.

Solution:

This is a chi squared (χ^2) test.

H_0: the coin is unbiased.

H_1: the coin is biased.

Under the null hypothesis the expected number of heads is 50 and tails is 50.

Draw up the following table:

	O	E	(O – E)	(O – E)²	$\frac{(O-E)^2}{E}$
Heads	60	50	10	100	2
Tails	40	50	-10	100	2

The χ^2 test statistic is $\chi^2 = \sum \dfrac{(O-E)^2}{E} = 4$

The degrees of freedom $v = 2 - 0 - 1 = 1$.

The 5% critical value with 1 degree of freedom is 3.84 from tables.

The χ^2 distribution can be represented on a diagram

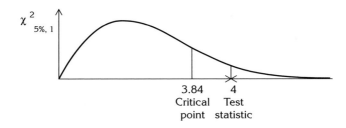

$\chi^2_{5\%,\,1}$

3.84   4
Critical   Test
point   statistic

Since our test statistic lies in the critical region we reject the null hypothesis, H_0, and conclude that the coin is biased.

critical value: see *critical region*.

cross product: the cross product (or vector product) of two vectors **a** and **b** is the vector **c** whose magnitude is $ab \sin \theta$ where θ is the angle between **a** and **b**. The direction of **c** is perpendicular to both **a** and **b** so that **a**, **b** and **c** form a right handed system.

$$|\mathbf{c}| = |\mathbf{a} \times \mathbf{b}| = ab \sin \theta$$

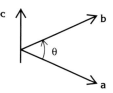

The cross product is an example of a mathematical operation which is not commutative:

$$\mathbf{b} \times \mathbf{a} = -\mathbf{a} \times \mathbf{b}$$

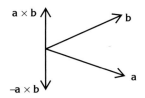

If **a** and **b** are given by Cartesian coordinates $\mathbf{a} = a_1\mathbf{i} + a_2\mathbf{j} + a_3\mathbf{k}$ and $\mathbf{b} = b_1\mathbf{i} + b_2\mathbf{j} + b_3\mathbf{k}$ then $\mathbf{a} \times \mathbf{b}$ is given by

$$\mathbf{a} \times \mathbf{b} = \begin{vmatrix} \mathbf{i} & \mathbf{j} & \mathbf{k} \\ a_1 & a_2 & a_3 \\ b_1 & b_2 & b_3 \end{vmatrix} = (a_2b_3 - a_3b_2)\mathbf{i} + (a_3b_1 - a_1b_3)\mathbf{j} + (a_1b_2 - a_2b_1)\mathbf{k}$$

A geometrical interpretation of the cross product is the magnitude of the area of the parallelogram formed by the vectors **a** and **b**.

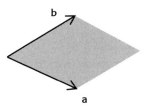

area $A = |\mathbf{a} \times \mathbf{b}|$

A unit vector that is perpendicular to the plane of the parallelogram is the vector

$$\mathbf{n} = \frac{\mathbf{a} \times \mathbf{b}}{|\mathbf{a} \times \mathbf{b}|}$$

Example:

Find a unit vector that is perpendicular to the two vectors:

$$\mathbf{a} = \mathbf{i} - 3\mathbf{j} + \mathbf{k} \quad \text{and} \quad \mathbf{b} = 2\mathbf{i} - \mathbf{j} + 3\mathbf{k}$$

Solution:

First we find the cross product **a** × **b**

$$\mathbf{a} \times \mathbf{b} = \begin{vmatrix} \mathbf{i} & \mathbf{j} & \mathbf{k} \\ 1 & -3 & 1 \\ 2 & -1 & 3 \end{vmatrix} = (-9 - (-1))\mathbf{i} + (2 - 3)\mathbf{j} + (-1 - (-6))\mathbf{k} = -8\mathbf{i} - \mathbf{j} + 5\mathbf{k}$$

The magnitude of **a** × **b** is given by

$$\sqrt{(-8)^2 + (-1)^2 + (5)^2} = \sqrt{90}$$

A unit vector in a direction that is perpendicular to both **a** and **b** is

$$\mathbf{e} = \frac{\mathbf{a} \times \mathbf{b}}{|\mathbf{a} \times \mathbf{b}|} = \frac{-8\mathbf{i} - \mathbf{j} + 5\mathbf{k}}{\sqrt{90}}$$

cube: this is a solid with six square faces, where adjacent faces are perpendicular.

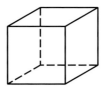

The term cube also applies to the operation of multiplying a number by itself three times. For example "five cubed," written as 5^3 is 125.

cube root: the cube root of a number is a number that when cubed gives the original number. For example the cube root of 8 is 2, because $2 \times 2 \times 2 = 8$. The cube root of x is usually written as $\sqrt[3]{x}$ or $x^{1/3}$.

cubic: this term usually describes polynomial equations or expressions that involve x^3 but no higher powers of x. For example the equation $x^3 - 5x^2 + 7x - 12 = 0$ is a "cubic equation." There is no simple method to solve cubic equations and when faced with an equation such as this it is often necessary to resort to a numerical method.

A cubic equation can have either one, two or three real *roots* as illustrated in the graphs below.

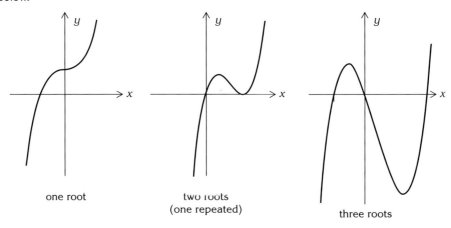

one root   two roots (one repeated)   three roots

A cubic with an equation of the form $y = A(x - a)^3$, will have only real root a. (Note that there are many different forms for the equation of a cubic with only one real root.)

A cubic with an equation of the form $y = A(x - a)^2(x - b)$, will have two real roots a and b. The root a is called a repeated root in these two cases.

A cubic with an equation of the form $y = A(x - a)(x - b)(x - c)$, will have three real roots, a, b and c.

cuboid: a solid similar in shape to a cube, but with faces that are rectangles rather than squares. As with a cube adjacent faces are perpendicular.

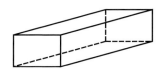

cumulative distribution function (c.d.f.): the cumulative distribution function is defined by

$$A = F(a) = \int_{-\infty}^{a} f(x)\, dx$$

where $f(x)$ is the *probability density function*, p.d.f.

$F(a) = P(X \leq a)$ and is the cumulative distribution function.

Example:

A continuous random variable X has a probability density function

$$f(x) = 3x^2 + 2x \quad \text{for } 0 \leq x \leq a$$
$$= 0 \text{ elsewhere}$$

Find the cumulative density function of X.

Solution:

Let $F(a) = \int_0^a f(x)\, dx$ be the cumulative distribution function of X.

Then $F(a) = \int_0^a 3x^2 + 2x\, dx = [x^3 + x^2]_0^a = a^2(1 + a)$

cumulative frequency: the total frequency of all values less than the upper class boundary of a given class interval, up to and including the class interval. For example for the data in the following table the cumulative frequency up to and including the class interval $69 - 71$ is $4 + 9 + 20 + 13 = 46$.

Mass (kg)	Number of students
60–62	4
63–65	9
66–68	20
69–71	13
72–74	4

A "cumulative frequency polygon" is the graph obtained when the cumulative frequency is plotted against the upper class boundary and the points joined with straight lines. If the points are joined with a smooth line the graph is known as the cumulative frequency curve or ogive. The cumulative frequency table for the data on student masses is shown below with its cumulative frequency polygon.

Upper class boundary	Frequency	Cumulative frequency
59.5	0	0
62.5	4	4
65.5	9	13
68.5	20	33
71.5	13	46
74.5	4	50

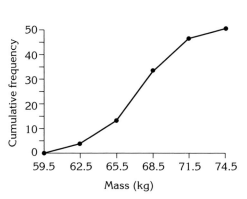

Percentage cumulative frequency is the scaled cumulative frequency. It is useful to use the percentage cumulative frequency when estimating the median and quartiles from the polygon or ogive.

Upper class boundary	Frequency	Cumulative frequency	Percentage cumulative frequency
59.5	0	0	0
62.5	4	4	8
65.5	9	13	26
68.5	20	33	66
71.5	13	46	92
74.5	4	50	100

curve sketching: when sketching the curve of a function $f(x)$, follow the procedure outlined below:

1. evaluate $f(0)$ to find where the curve crosses the y-axis
2. solve the equation $f(x) = 0$ to find where the curve crosses the x-axis
3. consider how the curve behaves as $x \to \pm\infty$
4. find any stationary points
5. If the function is of the form $\dfrac{p(x)}{q(x)}$ solve the equation $q(x) = 0$, to find the position of any discontinuities.

The curve can then be sketched.

Example:

Sketch the curve $y = \dfrac{x^2}{x - 2}$.

Solution:

Following the steps above.

1. If $x = 0$, then $y = \dfrac{0}{-2} = 0$, so the curve cuts the y-axis only at the point with coordinates $(0, 0)$.

2. Solving $\dfrac{x^2}{x - 2} = 0$ again shows that the curve passes through the point $(0, 0)$.

3. As $x \to \pm\infty$, $\dfrac{x^2}{x - 2} \to x$.

   In addition for large positive x, $\dfrac{x^2}{x - 2} > x$ and for large negative x, $\dfrac{x^2}{x - 2} < x$. So the line $y = x$ is an *asymptote* for large x, with the curve approaching this line from above for large positive x and from below for large negative x.

4. For this function

   $\dfrac{dy}{dx} = \dfrac{x(x - 4)}{(x - 2)^2}$ (using the *quotient rule*).

   So there are turning points at $x = 0$ and $x = 4$. Using the quotient rule again gives

   $\dfrac{d^2y}{dx^2} = \dfrac{8}{(x - 2)^3}$. When $x = 0$, $\dfrac{d^2y}{dx^2} = -1$

   so there is a local maximum at $(0, 0)$.

When $x = 4$, $\dfrac{d^2y}{dx^2} = 1$ so there is local minimum at $(4, 8)$.

5. The function will have a discontinuity when $x - 2 = 0$, that is $x = 2$. So the line $x = 2$ will be an asymptote. As x approaches 2 from below y becomes large and negative. As x approaches 2 from above y is large and positive.

With this information it is possible to sketch the curve. The above information is summarized on the diagram below.

The curve can then be sketched as shown below.

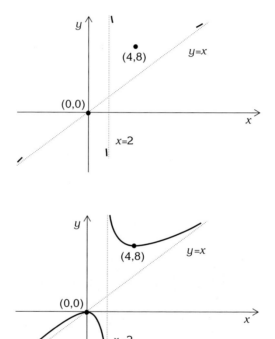

With a graphic calculator you would graph a function immediately and then use the steps 1–5 to explore the properties of the function.

cut: a line dividing a *network* into two parts, one containing the *source* and the other containing the *sink*. The overall value or capacity of a cut is the sum of the capacities of the *arcs* it crosses provided that the arcs are directed from the source to the sink.

cycle: a path that completes a loop and returns to its starting point.

cylinder: the volume of a cylinder of height h and base radius r is $\pi r^2 h$. The surface area of the curved surface of a cylinder is $2\pi rh$ and the total surface area of a closed cylinder is $2\pi r^2 + 2\pi rh = 2\pi r(r + h)$.

damping: the word used to describe a decaying motion. It is usually used in connection with oscillations and vibrations.

Any naturally oscillating system will decay with time. For example, the motion of a bungee jumper and a child on a swing gradually decay. We say that the oscillations are "damped."

Some oscillations decay very rapidly and such a system is said to be overdamped. If there are many oscillations before the oscillations become small then we say that the system is "underdamped."

The figure shows the suspension system for a car. It consists of a spring and a damper. It is carefully designed to provide a smooth ride over bumpy roads; the spring alone would provide a "bouncy ride" so the damper is included to damp down the vibrations.

The equation of motion of a damped system is

$$\frac{d^2x}{dt^2} + \alpha\frac{dx}{dt} + \beta x = 0$$

with $\alpha > 0$ and $\beta > 0$.

The classification of the system is summarized in the table on page 62.

Criteria	Classification	Graph of motion
$\alpha^2 - 4\beta > 0$	Overdamping	
$\alpha^2 - 4\beta = 0$	Critical damping	
$\alpha^2 - 4\beta < 0$	Underdamping	

deciles: the deciles, D_1, D_2, ..., D_9 divide a distribution into ten equal parts. D_1 has 10% of the distribution below it, D_2 has 20% below, etc.

decreasing function: a function is said to be a decreasing function on an interval if for any values a and b in the interval where $a < b$ we have $f(a) > f(b)$. Sometimes a function is described as being decreasing on an interval, (a, b) if $f'(x) < 0$ for $a < x < b$. For example $f(x) = 8 - x$ is a decreasing function because $f'(x) = -1$, which satisfies $f'(x) < 0$ for all values of x.

definite integral: this is an integral that is to be evaluated between two limits. For example

$$\int_2^5 3x^2 - 6x + 1\,dx$$

is a definite integral.

Example:

Find $\displaystyle\int_2^5 3x^2 - 6x + 1\,dx$

Solution:

$$\int_2^5 3x^2 - 6x + 1\,dx = [x^3 - 3x^2 + x]_2^5$$
$$= (5^3 - 3 \times 5^2 + 5) - (2^3 - 3 \times 2^2 + 2)$$
$$= 55 - (-2)$$
$$= 57$$

degree of a vertex: the number of *edges* incident at the *vertex*.

degrees of freedom (d.f.): can be considered the "number of free choices." For example, find three numbers, x, y, and z, which add up to 25. It is obvious that there are not three free choices of the numbers. Once two numbers have been chosen, the third number is fixed, i.e. $z = 25 - (x + y)$. Therefore, in this case we have three variables (x, y, and z) subject to one restriction ($x + y + z = 25$) and $3 - 1 = 2$ degrees of freedom.

In general, the number of degrees of freedom associated with a sample *test statistic* T:

$$\begin{matrix} \text{Number of degrees} \\ \text{of freedom} \end{matrix} = \begin{matrix} \text{Number of original} \\ \text{variables} \end{matrix} - \begin{matrix} \text{Number of restrictions involved} \\ \text{in the calculation of } T \end{matrix}$$

The degrees of freedom are usually denoted by the symbol v.

denominator: The denominator of a fraction is the number that appears on the bottom of the fraction. For example the denominator of a/b is b. (See also *numerator*.)

density: the average density of an object is the ratio of its mass to its volume:

$$\text{average density} = \frac{\text{mass}}{\text{volume}}$$

A fundamental assumption in modeling an object as a rigid body is that each subdivision of the rigid body has mass and volume. For each subdivision we can calculate the average density. The density of a body may change within the body as we move from subdivision to subdivision. However, if the density of the body is constant throughout then the body is called a *uniform* body.

density function: see *probability density function*.

derivative: the derivative of a function gives its rate of change or, for a curve, its gradient. The derivative of y with respect to x is written as

$$\frac{dy}{dx}$$

The derivative of $f(x)$ is written as $f'(x)$.

derivative of x^n: The derivative of x^n is nx^{n-1} for any value of n.

Example:

Differentiate each of the following.
(a) $y = 2x^7$

(b) $y = \dfrac{4}{x^3}$

(c) $y = \sqrt[4]{x}$

Solution:

(a) As $y = 2x^7$ differentiating gives

$$\frac{dy}{dx} = 2 \times 7x^{7-1} = 14x^6$$

(b) First note that $\dfrac{4}{x^3} = 4x^{-3}$. Then differentiating gives:

$$\frac{dy}{dx} = 4 \times (-3x^{-3-1}) = -12x^{-4} = \frac{-12}{x^4}$$

(c) Note that $\sqrt[4]{x} = x^{1/4}$. Then differentiating gives:

$$\frac{dy}{dx} = \frac{1}{4}x^{1/4-1} = \frac{1}{4}x^{-3/4} = \frac{1}{4} \times \frac{1}{\sqrt[4]{x^3}}$$

derivative of e^x: This function has the unique property that the derivative of e^x is simply e^x. The chain rule can be used to show that the derivative of e^{kx} is ke^{kx}.

Example:

Differentiate each of the following:

(a) $y = 4e^{7x}$

(b) $y = 5e^{-5x}$

(c) $y = \dfrac{5}{e^{2x}}$

Solution:

(a) $\dfrac{dy}{dx} = 4 \times 7e^{7x} = 28e^{7x}$

(b) $\dfrac{dy}{dx} = 5 \times (-5e^{-5x}) = -25e^{-5x}$

(c) First write $y = \dfrac{5}{e^{2x}}$ as $y = 5e^{-2x}$.

   Then differentiate to give:

   $\dfrac{dy}{dx} = 5 \times (-2e^{-2x}) = -10e^{-2x}$,

derivative of ln(x): The derivative of $\ln(x)$ is $1/x$ and the derivative of $\ln(kx)$ is also $1/x$. This second result can be shown using the chain rule. If $y = \ln(kx)$, differentiating gives

$$\frac{dy}{dx} = \frac{1}{kx} \times k = \frac{1}{x}.$$

derivatives of inverse trigonometric functions: the derivatives of the inverse trigonometric functions are listed in the table on the next page.

Function	Derivative
$\cos^{-1}(kx)$ or $\arccos(kx)$	$\dfrac{-k}{\sqrt{1 - (kx)^2}}$
$\sin^{-1}(kx)$ or $\arcsin(kx)$	$\dfrac{k}{\sqrt{1 - (kx)^2}}$
$\tan^{-1}(kx)$ or $\arctan(kx)$	$\dfrac{k}{1 + (kx)^2}$

The entry for $\cos^{-1}$ shows how these results are obtained.

Example:

Differentiate

$$y = 5 \tan^{-1}\left(\frac{x}{2}\right)$$

Solution:

$$\frac{dy}{dx} = 5 \times \frac{1/2}{1 + (x/2)^2} = 5 \times \frac{1/2}{1 + x^2/4} = 5 \times \frac{2}{4 + x^2} = \frac{10}{4 + x^2}$$

derivatives of trigonometric functions: these are needed when you are differentiating expressions that contain trigonometric functions to find gradients or rates of change. The table below provides a reference list of the derivatives of all the trigonometric functions.

Function	Derivative
$\sin(kx)$	$k\cos(kx)$
$\cos(kx)$	$-k\sin(kx)$
$\tan(kx)$	$k\sec^2(kx)$
$\sec(kx)$	$k\sec(kx)\tan(kx)$
$\operatorname{cosec}(kx)$	$-k\operatorname{cosec}(kx)\cot(kx)$
$\cot(kx)$	$-k\operatorname{cosec}^2(kx)$

Example:

Differentiate each of the following:
(a) $y = 5\sin(9x)$
(b) $y = 3\tan(2x)$
(c) $y = 3\cot(4x)$

Solution:

Using the rules to differentiate gives:

(a) $\dfrac{dy}{dx} = 5 \times 9\cos(9x) = 45\cos(9x)$

(b) $\dfrac{dy}{dx} = 3 \times 2\sec^2(2x) = 6\sec^2(2x)$

(c) $\dfrac{dy}{dx} = 3 \times (-4\operatorname{cosec}^2(4x)) = -12\operatorname{cosec}^2(4x)$

deterministic model: a model that assumes chance events do not occur.

difference of squares: the term difference of squares is often used to describe the factorization of $x^2 - y^2$, which gives $(x + y)(x - y)$.

Example:

Factorize:
(a) $x^2 - 16$
(b $4x^2 - 9y^2$
(c) $x^2 - 5$

Solution:

(a) $\quad x^2 - 16 = x^2 - 4^2$
$\qquad\qquad = (x + 4)(x - 4)$
(b) $4x^2 - 9y^2 = (2x)^2 - (3y)^2$
$\qquad\qquad = (2x + 3y)(2x - 3y)$
(c) $\quad x^2 - 5 = x^2 - (\sqrt{5})^2$
$\qquad\qquad = (x + \sqrt{5})(x - \sqrt{5})$

differential equation: A differential equation is an equation which contains an independent variable, a dependent variable and derivatives of the dependent variable with respect to the independent variable. If there is only the first derivative present in the equation then it is called a "first order differential equation." The most general form of a first order differential equation in the x–y variables is written as:

$$\frac{dy}{dx} = f(x, y)$$

The *general solution* of a differential equation contains arbitrary constants. If initial conditions and/or boundary equations are given, they can be used to find the *particular solution*. For a first order differential equation there is one arbitrary constant; for a second order differential equation there are two arbitrary constants and so on.

Differential equations are used as models to describe real situations. For example:

$$\frac{dP}{dt} = aP(M - P)$$

models the population growth of a species (called the Logistic equation)

$$\frac{dv}{dt} = mF(v, t)$$

models the motion of a particle of mass m subject to a force F

$$\frac{dT}{dt} = -a(T - T_0)$$

models the cooling of an object (called Newton's law of cooling).

Example:

Solve the differential equation $x\dfrac{dy}{dx} - y = xy$, given that $y(2) = 1$.

Solution:

The differential equation is first order and can be written in standard form

$$\frac{dy}{dx} = \frac{(1 + x)y}{x}$$

Separating the variables and integrating

$$\int \frac{1}{y}\,dy = \int \frac{(1 + x)}{x}\,dx = \int \left(\frac{1}{x} + 1\right)dx$$

$$\ln |y| = \ln |x| + x + c$$

Substituting for $x = 2$, $y = 1$

$$\ln (1) = \ln (2) + 2 + c \Rightarrow c = -\ln (2) - 2$$

Therefore

$$\ln |y| = \ln |x| + x - \ln (2) - 2$$

$$y = e^{\ln |x| + x - \ln (2) - 2}$$

$$= e^{\ln |x| - \ln (2)}\, e^{x - 2}$$

$$= \frac{x}{2} e^{(x - 2)}$$

differentiation: the process of finding the derivative of a function.

differentiation from first principles: the formula below can be used to find the derivative of a function from first principles.

If $y = f(x)$ then

$$\frac{dy}{dx} = \lim_{h \to 0} \left(\frac{f(x + h) - f(x)}{h} \right)$$

This result is obtained by considering the gradient of a chord between two points on the curve $y = f(x)$.

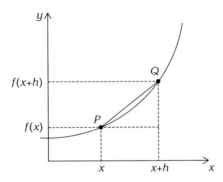

The gradient of the chord PQ is given by

$$\frac{f(x + h) - f(x)}{h}$$

As Q approaches P the gradient of the chord becomes closer and closer to the gradient of the tangent line to the curve at P. The limit of the gradient of the chord as h tends to 0 is called the gradient of the curve at x and denoted by

$$\frac{dy}{dx} = \lim_{h \to 0} \left(\frac{f(x + h) - f(x)}{h} \right).$$

Example:

Find the derivative of x^3 from first principles.

Solution:

Using $f(x) = x^3$ gives

$$\frac{dy}{dx} = \lim_{h \to 0} \left(\frac{f(x + h) - f(x)}{h} \right)$$

$$= \lim_{h \to 0} \left(\frac{(x + h)^3 - x^3}{h} \right)$$

$$= \lim_{h \to 0} \left(\frac{x^3 + 3x^2h + 3xh^2 + h^3 - x^3}{h} \right)$$

$$= \lim_{h \to 0} \left(\frac{3x^2h + 3xh^2 + h^3}{h} \right)$$

67

$$= \lim_{h \to 0} (3x^2 + 3xh + h^2)$$

$$= 3x^2$$

Dijkstra's algorithm: an example of a *greedy algorithm* for finding the shortest route between two *vertices* in a *network*.

Step 1: assign the permanent label 0 to the starting vertex.

Step 2: assign temporary labels to all the vertices that are connected directly to the most recently permanently labeled vertex.

Step 3: choose the vertex with the smallest temporary label and assign a permanent label to that vertex.

Step 4: repeat steps 2 and 3 until all vertices have permanent labels.

Step 5: find the shortest path by tracing back through the *network*.

dimensions: all quantities in mechanics can be expressed in terms of the three fundamental quantities: mass, length and time. The dimensions of a quantity show how it is related to these three fundamental quantities.

The notation used for the dimensions are M for mass, L for length and T for time. A square bracket notation is used, for example, $[v]$ means the dimension of speed.

The dimensions of a quantity are independent of the units used. So for example the dimensions of speed are LT^{-1} irrespective of whether we are using mph or m s^{-1}.

The following table shows the dimensions of common quantities in mechanics.

quantity	dimensions
position	L
velocity	LT^{-1}
acceleration	LT^{-2}
force	MLT^{-2}
energy	ML^2T^{-2}
work	ML^2T^{-2}
angle	dimensionless
coefficient of friction	dimensionless
stiffness of a spring	MT^{-2}

To find the dimensions of a quantity we use an equation that defines or uses the quantity.

Example:

Find the dimensions of

(a) area

(b) the constant in the quadratic model for air resistance $R = kv^2$ where R is the magnitude of the air resistance at speed v.

Solution:

(a) Area is the product of two lengths, for example, for a rectangle it is length ×
breadth $[A] = L^2$

(b) For the model of air resistance $R = kv^2$, $[R] = MLT^{-2}$ and $[v] = LT^{-1}$, so

$$[R] = [k][v^2]$$

$$[k] = \frac{[R]}{[v^2]} = \frac{MLT^{-2}}{L^2T^{-2}} = ML^{-1}$$

---

dimensional consistency: the two sides of an equation that model a situation in
mechanics must have the same dimensions. We say that the equation is "dimensionally
consistent" or "dimensionally homogenous."

Example:

The following models are proposed for the velocity of a particle as a function of
time. Which equation could not be a good model?

(a) $v = u + at$
(b) $v^2 = u^2 + at^2$

where u is the initial velocity and a is the acceleration.

Solution:

We check the dimensions of each side of the equation:

(a) $[v] = LT^{-1}$, $[u] = LT^{-1}$, $[a] = LT^{-2}$, $[t] = T$
$[at] = LT^{-2} \times T = LT^{-1}$

Each term in the equation $v = u + at$ has the same dimensions so the equation is
dimensionally consistent. This means that it could be a model for a mechanics
situation.

(b) $[v^2] = L^2T^{-2}$, $[u^2] = L^2T^{-2}$, $[at^2] = LT^{-2} \times T^2 = L$
The dimensions of the third term are different from the other two terms in the
equation, so the equation is not dimensionally consistent. This equation cannot be
a model for a mechanics situation.

Example:

A model for the magnitude of the air resistance force on a ball bearing falling through
a liquid is proposed as: $R = km^a r^b v^c$; where a, b and c are numbers, k is a dimensionless constant, m and r are the mass and radius of the ball bearing and v is its
speed. Find values of a, b and c for the model to be dimensionally consistent.

Solution:

The dimensions of R are MLT^{-2} so

$$MLT^{-2} = [k] \times [m^a] \times [r^b] \times [v^c] = M^a L^b (LT^{-1})^c = M^a L^{b+c} T^{-c}$$

For dimensional consistency

$a = 1$ $b + c = 1$ $c = 2$

69

So $a = 1$, $b = -1$ and $c = 2$, which gives the possible model

$R = kmr^{-1}v^2$

---

direct impact: a direct impact of two bodies occurs when both of the bodies have a velocity that is along a common normal.

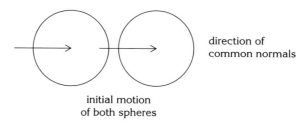

direction of
common normals

initial motion
of both spheres

(See *coefficient of restitution*.)

directed graph: a *graph* with directions on some or all of the *edges*.

directed network: a *network* in which there is a set direction of movement along some or all of the *edges*.

direction field: this is constructed from a differential equation to give information about the solutions of the differential equation. In a direction field, short lines are plotted that would have the same gradient as the solution curve that passes through the point at the center of the line. The direction field below is for the differential equation

$$\frac{dy}{dx} = -\frac{x}{y}$$

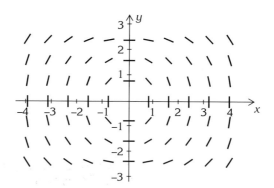

The direction field shows that the solutions can be represented as circles. In fact the general solution of the differential equation is $x^2 + y^2 = c$, with the value of c determining the radius of the circle.

discontinuity: some curves are continuous and have no breaks in them, but others with a break are said to "have a discontinuity." The graph on the next page shows

$$y = \frac{x}{x-1}$$

which has a discontinuity at $x = 1$.

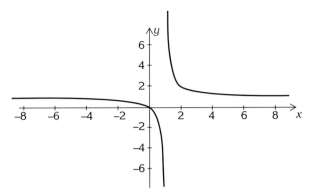

A discontinuity

discrete data: a type of *quantitative data*. They can take only known quantities which can be identified exactly. The number of rooms in a house, the number of people in a class, shoe sizes and the number of letters in a surname are all examples of discrete data. (See also *continuous data*.)

discrete probability distribution: a discrete probability distribution is the probability distribution of a discrete random variable, X, which can take only the discrete values $x_1, x_2, \ldots, x_n$ with probabilities $p_1, p_2, \ldots, p_n$ respectively. (See also *discrete random variable*, *probability density function*.)

discrete random variable: if X is a discrete variable taking only the values $x_1, x_2, \ldots x_n$ with probabilities $p_1, p_2, \ldots p_n$ respectively, and

$$\sum_{i=1}^{n} p_i = 1$$

then X is a discrete random variable.

Example:

A discrete variable X takes on the values $x_1, x_2, \ldots$, with probability $p_1, p_2, \ldots$, as shown.

x	0	1	2	3	4
$P(X = x)$	0.2	0.3	0.25	0.15	0.1

Show that X is a discrete random variable.

Solution:

$$\sum_{i=1}^{n} p_i = 0.2 + 0.3 + 0.25 + 0.15 + 0.1 = 1$$

and therefore X is a discrete random variable.

discrete situation: a situation modeled by a discrete sequence of numbers.

discrete variable: see *discrete random variable*.

discriminant: the discriminant of the quadratic equation $ax^2 + bx + c = 0$ is $b^2 - 4ac$. The discriminant determines how many real roots the equation has.

(See also *quadratic equations*.)

dispersion (measures of): a measure of dispersion is a numerical quantity that indicates how scattered or spread the data is. Such measures include *range*, *interquartile range*, *mean deviation*, *standard deviation* and *variance*.

displacement: the displacement of an object at a point B relative to a point A is the difference between the position vectors of B and A relative to a fixed origin.

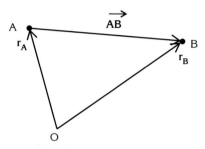

Consider an object moving along any path between A and B then the displacement vector is:

$$\overrightarrow{AB} = r_B - r_A$$

The displacement should not be confused with the actual distance traveled between A and B.

Example:

An object moves in a circle of radius 0.7 m. Find (a) the displacement of the object and (b) the distance traveled as the object moves from A to (i) B, (ii) C and (iii) D.

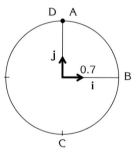

Solution:

The position vectors of A, B, C and D are

$$r_A = 0.7\,j \qquad r_B = 0.7\,i \qquad r_C = -0.7\,j \qquad r_D = 0.7\,j$$

(a) (i)   the displacement of B relative to A = $0.7\,i - 0.7\,j$
   (ii) the displacement of C relative to A = $-1.4\,j$
   (iii) the displacement of D relative to A = **0**

(b) the distances traveled are:
   (i) to B, $0.7 \times (\pi/2)$;(ii) to C, $0.7 \times \pi$; (iii) to D, $1.4 \times \pi$

For motion in a straight line the displacement is often the position of the object relative to the origin and is denoted by x.

The value of x is positive to the right of the origin and negative to the left. The magnitude of the displacement is then the distance from the origin $|x|$.

A graph that illustrates the displacement at various times is called a *displacement–time graph*. The figure below shows the displacement-time graph for an object that initially moves to the left, then stops, then moves to the right along a straight line.

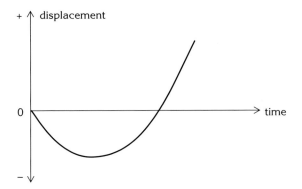

Sometimes these graphs are called distance–time graphs, which is not correct. It is important not to confuse displacement with distance traveled.

Example:

Draw a distance–time graph for the object above.

Solution:

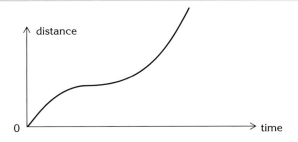

disproof by counterexample: can be used to show that a statement or conjecture is not true by providing one example of a case where the statement or conjecture is not true.

For example, consider the statement $x^2 > x$. It can be shown that this statement is not true by considering the case of $x = \dfrac{1}{2}$. In this case we would have:

$$\left(\frac{1}{2}\right)^2 > \frac{1}{2} \quad \text{or} \quad \frac{1}{4} > \frac{1}{2},$$

so the statement is clearly not true.

distance matrix: a *cost matrix* in which the numbers represent distances between *vertices*.

distribution: see *binomial, bivariate, chi squared, frequency, normal, Poisson, probability.*

distributive: an operation $\otimes$ is said to be distributive over another operation $*$, if $a \otimes (b * c) = (a \otimes b)*(a \otimes c)$. For example, multiplication is distributive over addition because $a(b + c) = ab + ac$. However the same is not true if multiplication is replaced by division because $a \div (b + c) \neq (a \div b) + (a \div c)$.

division of complex numbers: can be carried out by multiplying both numbers by the conjugate of the divisor. This produces a divisor that is not complex and allows the division to take place.

The diagram below shows how the division of complex numbers can be treated geometrically.

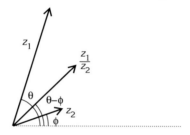

Note that

$$\left|\frac{z_1}{z_2}\right| = \frac{|z_1|}{|z_2|} \quad \text{and} \quad \arg\left(\frac{z_1}{z_2}\right) = \arg(z_1) - \arg(z_2)$$

Example:

Find $\dfrac{3 + 2i}{5 - 3i}$.

Solution:

First multiply the top and bottom by the conjugate of $5 - 3i$, which is $5 + 3i$. Then simplify the result as shown below.

$$\frac{3 + 2i}{5 - 3i} = \frac{(3 + 2i)(5 + 3i)}{(5 - 3i)(5 + 3i)}$$

$$= \frac{15 + 9i + 10i + 6i^2}{25 + 15i - 15i - 9i^2}$$

$$= \frac{9 + 19i}{34}$$

$$= \frac{9}{34} + \frac{19i}{34}$$

division of polynomials: To divide a polynomial by another polynomial the usual rules of algebra are used.

Example:

Divide $7x^3 - 6x^2 + x - 3$ by $x^2 - x + 2$

Solution:

$$
\begin{array}{r}
7x + 1 \\
x^2 - x + 2 \overline{)\, 7x^3 - 6x^2 + x - 3} \\
7x^3 - 7x^2 + 14x \\
\hline
x^2 - 13x - 3 \\
x^2 - x + 2 \\
\hline
-12x - 5
\end{array}
$$

Hence $(7x^3 - 6x^2 + x - 3) \div (x^2 - x + 2) = 7x + 1 - \dfrac{(12x + 5)}{x^2 - x + 2}$

division of surds: see *surds*.

domain: the domain of a function is a set of numbers to which the function is applied. For more details see *function*.

dot product: the dot product (or *scalar product*) of two vectors **a** and **b** is the scalar quantity defined by

$$\mathbf{a}.\mathbf{b} = |\mathbf{a}|\,|\mathbf{b}|\,\cos\theta$$

where θ (for $0 \le \theta \le \pi$) is the angle between the directions of the two vectors.

If **a** and **b** are given by Cartesian coordinates $\mathbf{a} = a_1\mathbf{i} + a_2\mathbf{j} + a_3\mathbf{k}$ and $\mathbf{b} = b_1\mathbf{i} + b_2\mathbf{j} + b_3\mathbf{k}$, then $\mathbf{a}\cdot\mathbf{b}$ is given by

$$\mathbf{a}.\mathbf{b} = a_1 b_1 + a_2 b_2 + a_3 b_3$$

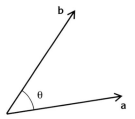

An important consequence of the definition of the dot product is that if **a** and **b** are two non-zero vectors and $\mathbf{a}\cdot\mathbf{b} = 0$ then $\cos\theta = 0$ which implies that the two vectors are perpendicular.

The dot product is often used to find the angle between two vectors.

Example:

Find the angle between the two vectors $\mathbf{a} = \mathbf{i} + 2\mathbf{j} + \mathbf{k}$ and $\mathbf{b} = -2\mathbf{i} + 4\mathbf{j} - \mathbf{k}$

Solution:

If θ is the angle between **a** and **b** then $\cos \theta = \dfrac{\mathbf{a} \cdot \mathbf{b}}{ab}$

For the two given vectors:

$$\mathbf{a} \cdot \mathbf{b} = a_1 b_1 + a_2 b_2 + a_3 b_3 = -2 + 8 - 1 = 5$$
$$a = |\mathbf{a}| = \sqrt{(1^2 + 2^2 + 1^2)} = \sqrt{6}$$
$$b = |\mathbf{b}| = \sqrt{(2^2 + 4^2 + 1^2)} = \sqrt{21}$$

Hence $\cos 0 = \dfrac{5}{\sqrt{6}\sqrt{21}} = 0.4454 \Rightarrow 0 = 63.55°$

double angle formulas: the double angle formulas are listed below:

$$\sin (2A) = 2 \sin A \cos A$$
$$\cos (2A) = \cos^2 A - \sin^2 A$$
$$= 2 \cos^2 A - 1$$
$$= 1 - 2 \sin^2 A$$
$$\tan (2A) = \frac{2 \tan A}{1 - \tan^2 A}$$

The double angle formulas are derived from the appropriate sum formulas. For example:

$$\sin (2A) = \sin (A + A)$$
$$= \sin A \cos A + \cos A \sin A$$
$$= 2 \sin A \cos A$$

The double angle formulas can be used to simplify trigonometric expressions, which is often particularly helpful for some integration problems.

Example:

Find $\int \cos^2 x \, dx$

Solution:

First rearrange the double angle formula

$$\cos (2x) = 2 \cos^2 x - 1$$

to give

$$\cos^2 x = \frac{1}{2} \cos (2x) + \frac{1}{2}$$

This can be used to simplify the integral:

$$\int \cos^2 x \, dx = \int \frac{1}{2} \cos (2x) + \frac{1}{2} \, dx$$

$$= \frac{1}{4} \sin (2x) + \frac{1}{2} x + c$$

dummy arc: an *arc* introduced when an activity has to be preceded by more than one other activity; dummy arcs are marked as dotted lines and have zero duration.

dynamic programming: a method of working backwards to solve *shortest path problems*. It has the advantage of being able to deal with negative edge weights. The algorithm works by evaluating the shortest path at each stage of the problem.

dynamical system: refers to an object, or set of objects, the motion of which we are investigating.

A simple pendulum swinging back and forth, the planets in the solar system, and the atmosphere near the Earth's surface are all examples of dynamical systems. But systems do not have to be physical systems. For example, population modeling in biology, the study of lead pollution in the environment, and models from those of the stock market or world economy to those describing chemical reactions involve dynamical systems.

The essential feature of a dynamical system is that its configuration or state is changing.

dynamics: the study of an object or system of objects that moves. Dynamics is part of the area of applied mathematics called mechanics. The study of mechanics in the context of an object or system of objects that remains at rest is called "statics."

edge: an edge is a line along which two faces of a solid or polygon meet. The diagram below shows a tetrahedron. Its edges are the six lines AB, AC, AD, BC, BD and CD. A cube has 12 edges.

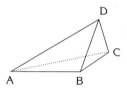

edge (arc): a line connecting two *vertices* in a *graph*.

edge-disjoint: different paths between two vertices are edge-disjoint if they have no edges in common.

elastic constant: see *stiffness*.

elastic impact: when two objects collide they either separate and remain as two objects or they coalesce and become one object. If they separate then the collision is called an *elastic collision*; otherwise it is an *inelastic collision*.

The type of collision depends on the material of the objects. The "bouncyness" or elasticity of the collision is measured by a number called the *coefficient of restitution*, usually denoted by the symbol *e*, which is a measure of the loss of kinetic energy in the collision. If no energy is lost the collision is called *perfectly elastic* and *e* = 1. For a perfectly inelastic collision *e* = 0. Generally the value of *e* for a collision between two objects lies between 0 and 1.

For example, if two billiard balls collide then they bounce off each other and there is little loss in kinetic energy in the collision; in modeling the collision of billiard balls we usually assume a perfectly elastic collision. If two lumps of soft putty collide they are likely to coalesce in an inelastic collision.

elastic potential energy: this is the potential energy stored in an elastic string when it is stretched or in an elastic spring when it is stretched or compressed. If the stiffness of the string or spring is *k* and it is stretched (or compressed) from its natural length by an amount *x* meters, then the elastic potential energy stored in the string or spring is ½ kx^2 (measured in joules).

The elastic potential energy provides a useful means of problem solving in mechanics.

Example:

A bungee jumper of mass 60 kg chooses a bungee rope of natural length 20 meters and stiffness 480 N m⁻¹.
What minimum height can she drop from and not hit the ground?

Solution:

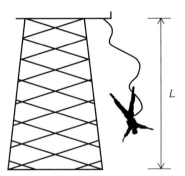

At the top of the jump the bungee jumper is at rest with a slack rope and at the bottom of the jump she is at instantaneous rest (before shooting back upwards!) with the rope stretched. So the gravitational potential energy at the top of the jump is converted into elastic potential energy at the bottom of the jump.

Model the bungee jumper as a particle and let the height of the jump be *L*. Then

$$mgL = \tfrac{1}{2}k(L-20)^2$$
$$60 \times 9.8 \times L = \tfrac{1}{2} \times 480 \times (L-20)^2$$

Solving for *L*, we get *L* = 28.33

The bungee jumper falls a distance of 28.33 meters before instantaneously stopping and then traveling upwards.

elasticity, modulus of: see *stiffness*.

elimination method: this is used to solve linear simultaneous equations. See *simultaneous equations* for more details.

elimination stage: the stage of the *Gaussian elimination method* involving the process of subtracting or adding a multiple of one row from another.

ellipse: the diagram below shows an ellipse centered on the origin. Its equation is

$$\frac{x^2}{a^2} + \frac{y^2}{b^2} = 1$$

In parametric form its equations are $x = a \cos t$ and $y = b \sin t$.

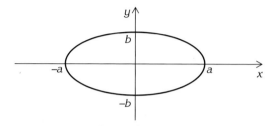

energy: a scalar quantity associated with an object or a system which occurs in different forms: heat, light, sound are common examples of energy. In mechanics the two forms of energy that concern us are *kinetic energy* and *potential energy*. These forms of energy are called mechanical energy.

Kinetic energy is the energy of an object associated with its speed or velocity and is defined by $\frac{1}{2}mv^2$.

Potential energy is the energy of an object associated with its position or configuration. For example the gravitational potential energy of an object of mass m at a height h above the level of zero potential energy is mgh. The elastic potential energy of a stretched elastic string of stiffness k and extension x is $\frac{1}{2}kx^2$.

The unit of energy is the joule (symbol J).

$$1 \text{ J} = 1 \text{ kg m}^2\text{s}^{-2}$$

Example:

Describe the energy changes that take place during a pole vault.

Solution:

As the athlete runs along the track chemical energy is turned into kinetic energy (and probably some heat and sound). During the vault the kinetic energy is turned into gravitational potential energy and elastic potential energy as the pole is bent. From the top of the vault the athlete gains kinetic energy losing gravitational potential energy.

(See also *elastic potential energy*.)

energy conservation: see *conservation of energy*.

equation of a straight line: the equation of a straight line is $y = mx + c$, where m is the gradient and c is the intercept with the vertical axis.

The equation of the line that passes through the two points with coordinates (x_1, y_1) and (x_2, y_2) can be found using

$$\frac{y - y_1}{y_2 - y_1} = \frac{x - x_1}{x_2 - x_1}$$

rather than finding the gradient and intercept separately.

equilateral: a shape is described as being equilateral if the lengths of all its sides are the same. An equilateral triangle has three sides that are all the same length.

equilibrium: an object or system of objects is said to be "in equilibrium" if there is no change in motion. This means that an object is in equilibrium if:

- it is at rest and remains at rest; this is called *static equilibrium*
- it moves with constant velocity; this is called *dynamic equilibrium*.

Note that constant velocity means moving in a straight line with constant speed.

In both cases the net (or resultant) force and net (or resultant) moment of the forces on the object, are zero.

When an object is in equilibrium we often say that "the forces are in equilibrium" or "in balance."

Example:

For each of the following situations decide if the object is in equilibrium.

(a) a crate on the back of a truck
(b) a bungee jumper
(c) a pebble dropped into a deep pool
(d) a car parked on a hill.

Solution:

(a) The crate will be in equilibrium if it is at rest or moving with constant velocity. The crate will be in equilibrium if it is at rest relative to the truck and the truck is at rest or moves with constant velocity (i.e. along a straight road with constant speed).
(b) The bungee jumper will be in equilibrium only when waiting to jump. During the down and up motions of the jump the jumper will not be in equilibrium. She is always accelerating or decelerating.

(c) A pebble dropped into a pool is not in equilibrium until it comes to rest on the bottom of the pool or it reaches its terminal speed.

(d) A car parked on a hill is in equilibrium.

errors (type I errors, type II errors): in *hypothesis testing*, a "type I error" is made when the null hypothesis H_0 is rejected when it should have been accepted. The probability of a "type I error" occurring, a, is the significance level of the test.

A type II error occurs when the null hypothesis H_0 is accepted when it should have been rejected. The probability of a type II error occurring, b, is more difficult to calculate.

estimation: a point estimate is used to find an estimator for the population parameter from the *sample parameter*.

An unbiased estimator is a sample statistic with *expectation* equal to the population parameter.

The best estimator is the unbiased estimator that has the smallest variance.

The *mean*, $\bar{x}$, of a sample can be used to find the estimated mean, μ, of the population. The best estimator for μ is $\bar{x}$.

The sample *variance*, s^2, can be used to find the population variance, σ^2. The best estimator for σ^2 is

$$\frac{ns^2}{n-1}$$

where n is the number in the sample.

The *proportion*, r, of a sample can be used to find the estimated proportion, Π, of the population. The best estimator for Π is r.

An interval estimate is used to find an interval for an unknown population parameter from the data from a random sample. Such intervals are *confidence intervals*.

Example:

The test scores (percentages) of a random sample of ten students from a class of 40 were recorded as below:

47   53   72   61   46   80   59   55   64   43

Find the best estimator for the population mean score and variance. Construct a 95% confidence interval for the population mean score assuming that the scores are normally distributed.

Solution:

The mean $\bar{x}$, of the sample is 58%. Therefore the best estimator of the population mean, μ, is 58%.

The variance, s^2, of the data is 125. Therefore the best estimator of the population variance, σ^2, is

$$\frac{ns^2}{n-1} = \frac{10 \times 125}{10-1} = 138.9$$

A 95% confidence interval for the mean μ of the population is given by

$$\bar{x} - z\,\frac{\sigma}{\sqrt{n}} \;<\; \mu \;<\; \bar{x} + z\,\frac{\sigma}{\sqrt{n}}$$

where z is the $\frac{1}{2}(100 - 95)\%$ point of a standard normal distribution = 1.96.

$$\bar{x} - z\,\frac{\sigma}{\sqrt{n}} \;<\; \mu \;<\; \bar{x} + z\,\frac{\sigma}{\sqrt{n}}$$

$$58 - 1.96 \times \frac{\sqrt{138.9}}{\sqrt{10}} \;<\; \mu \;<\; 58 + 1.96 \times \frac{\sqrt{138.9}}{\sqrt{10}}$$

$$50.695 \;<\; \mu \;<\; 65.305.$$

Therefore a 95% confidence interval for the mean of the test scores is

$$50.70 \;<\; \mu \;<\; 65.31.$$

Eulerian cycle: a cycle that includes every *edge* of a *graph* exactly once.

even function: a function $f(x)$ is an even function if $f(x) = f(-x)$ for all x in the domain of f. The graph of $y = f(x)$ for an even function has the y axis as a line of symmetry. Examples of even functions are $f(x) = x^2$ and $f(x) = \cos(x)$. The graphs of $y = f(x)$ for these functions are shown below. Note that the y axis is a line of symmetry.

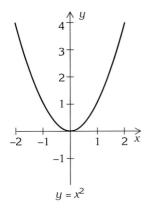

$y = x^2$

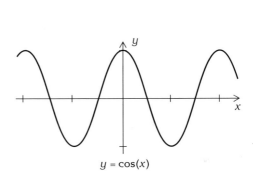

$y = \cos(x)$

(See *odd function*.)

even vertex: a *vertex* whose degree is even.

event: (1) an event in a *precedence network* is the finish of one activity and the start of another.

event: (2) an event, E, of an *experiment* is a subset of the *outcome set* S when probabilities are being considered. (See also *combined events*.)

Example:

A card is drawn from a standard pack of 52 cards. What is the probability that the card drawn is a heart?

Solution:

Let H be the event "a heart is drawn." Therefore the probability P of the event H occurring is given by $P(\text{H}) = \frac{1}{4}$.

exhaustive events: a set of events E_1, E_2, ... E_n of an experiment are said to be exhaustive events if their union is the outcome set, S. i.e. $E_1 \cup E_2 \cup ... \cup E_n = S$.

Example:

Consider the experiment rolling a die and the events E_1 the number is odd, and E_2 the number is even. Show that the events E_1 and E_2 are exhaustive events.

Solution:

Here, the outcome set

$$S = \{1, 2, 3, 4, 5, 6\}, \qquad E_1 = \{1,3,5\} \qquad and \qquad E_2 = \{2,4,6\}.$$
$$E_1 \cup E_2 = \{1, 2, 3, 4, 5, 6\} = S$$

Therefore the events E_1 and E_2 are exhaustive events.

exhaustive search: finding the shortest path by looking for every possible path.

expectation: given a random variable X which has a probability density function $P(X = x)$ for a discrete *random variable* or $f(x)$ for a *continuous random variable,* the expectation of X, $E[X]$, is given by:

$$E[X] = \begin{cases} \displaystyle\sum_{all\ x} xP(X = x) & \text{for discrete random variables} \\[2em] \displaystyle\int_{all\ x} x\,f(x)\ dx & \text{for continuous random variables} \end{cases}$$

Properties of $E[X]$:

$\qquad E[a] = a$ where a is a constant

$\qquad E[aX] = aE[X]$

$$E[G(X)] = \begin{cases} \displaystyle\sum_{all\ x} G(X)\ P(X = x) & \text{for discrete random variables} \\[2em] \displaystyle\int_{all\ x} G(X)\ f(x)\ dx & \text{for continuous random variables} \end{cases}$$

$\qquad E[F(X) + G(X)] = E[F(X)] + E\ [G(X)]$, where $F(X)$ and $G(X)$ are functions of X.

$\qquad E[aX + bY] = aE[X] + bE[Y]$, where a and b are constants and X and Y are any two random variables.

Example:

A discrete random variable X has the probability density function:

x	0	1	2	3	4
$P(X = x)$	0.2	0.3	0.25	0.15	0.1

and the discrete random variable Y has the probability density function:

y	2	3	4	5	6
$P(Y = y)$	0.1	0.25	0.3	0.2	0.15

Find (a) E[X]   (b) E[Y]   (c) E[3X]   (d) E[4Y]   (e) E[3X +4Y]

Solution:

(a) $E[X] = \sum_{\text{all } x} xP(X = x)$

$= (0 \times 0.2) + (1 \times 0.3) + (2 \times 0.25) + (3 \times 0.15) + (4 \times 0.1)$

$= 0 + 0.3 + 0.5 + 0.45 + 0.4$

$= 1.65$

(b) $E[Y] = \sum_{\text{all } y} yP(Y = y)$

$= (2 \times 0.1) + (3 \times 0.25) + (4 \times 0.3) + (5 \times 0.2) + (6 \times 0.15)$

$= 0.2 + 0.75 + 1.2 + 1.0 + 0.9$

$= 4.05$

(c) $E[3X] = 3E[X]$

$= 3 \times 1.65$

$= 4.95$

(d) $E[4Y] = 4E[Y]$

$= 4 \times 4.05$

$= 16.2$

(e) $E[3X + 4Y] = 3E[X] + 4E[Y]$

$= 4.95 + 16.2$

$= 21.15$

Example:

A continuous random variable has the probability density function $f(x) = k(x^2 + 3x)$ for $0 \le x \le 3$. Find (a) E[X] (b) E[3X]

Solution:

(a) $E[X] = \int_0^3 x f(x)\, dx$

$= k \int_0^3 x(x^2 + 3x)\, dx$

$= k \int_0^3 x^3 + 3x^2\, dx$

$= k \left[\dfrac{x^4}{4} + x^3 \right]_0^3$

$= 47.25k$

(b) $E[3X] = 3E[X]$

$= 3 \times 47.25k$

$= 141.75k$

experimental error: in the study of a possible relationship between two variables x and y, the independent variable, x, can be measured (or set) with little or no error, while the dependent variable, y, is subject to random experimental error. Such errors may be due to random fluctuations in experimental conditions and/or limitations in experimental apparatus.

experimental laws: relating two variables can be determined by using *lines of best fit* or *logarithmic plots*.

exponent: see *indices*.

exponential decay: the decay of a quantity that decreases according to the exponential function. The general shape of the decaying quantity is shown in the following figure:

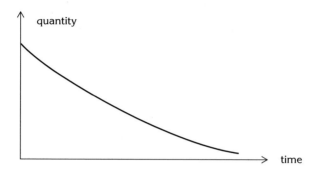

If the rate of decay of a quantity is exponential and an initial amount decreases by a constant percentage in a fixed time interval, then the quantity decreases by the same percentage in each subsequent time interval.

Radioactive elements decay according to this model. For example, the number of atoms in the element radium reduces by 50% approximately every 1600 years.

The time interval for a 50% reduction in an exponential decay model is called the "half-life."

exponential equation: equations which involve unknown powers or indices are exponential equations. For example $3 + 2^x = 6$ is an exponential equation. To solve such equations, the term containing the unknown power should first be made the subject of the equation, and then logs can be taken to produce a linear equation.

Example:

Solve the equation $3 + 2^x = 6$

Solution:

First make 2^x the subject of the equation:

$$3 + 2^x = 6$$
$$2^x = 3$$

Then take logs of both sides of the equation:

$$\log (2^x) = \log (3)$$
$$x \log (2) = \log (3)$$

$$x = \frac{\log (3)}{\log (2)}$$

$$= 1.58 \text{ (to 2 decimal places)}$$

exponential function: such a function has the form $f(x) = a^x$, where a is a nonnegative constant number and x is the independent variable.

e^x is called the natural exponential function where e is the *irrational number* 2.718281828 correct to ten significant figures. It is often shown on a calculator as "exp" and the exponential function is often written as $\exp(x)$.

The choice of e leads to the following properties for e^x:

● the gradient function (i.e. the derivative) of e^x is e^x

● e^x can be defined as the following limit:

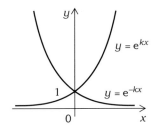

$$e^x = \lim_{n \to \infty} \left(1 + \frac{x}{n}\right)^n$$

● the series expansion of e^x is

$$1 + x + \frac{x^2}{2!} + \frac{x^3}{3!} + \frac{x^4}{4!} + \dots$$

The function e^{kx} or $\exp(kx)$ models *exponential growth* if $k > 0$ and *exponential decay* if $k < 0$.

Graphs of $y = e^{kx}$ for $k > 0$ and $k < 0$ are shown in the following diagram:

$$y = e^{kx}$$

$$y = e^{-kx}$$

exponential growth: the growth of a quantity that increases according to the exponential function. The general shape of the growing quantity is shown in the following figure:

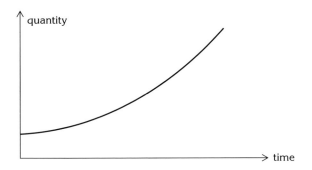

If the rate of growth of a quantity is exponential and an initial amount increases by a constant percentage in a fixed time interval, then the quantity increases by the same percentage in each subsequent time interval.

exponential series

For example, suppose that you invest $1000 in a savings account which gives an annual rate of interest of 4%, then the investment will grow at 4% in each time interval of one year.

The formula for the value of the investment after n years is:

$$I = 1000(1 + 0.04)^n = 1000 \times 1.04^n$$

exponential series: the series

$$1 + x + \frac{x^2}{2!} + \frac{x^3}{3!} + \frac{x^4}{4!} + \ldots$$

converges to e^x. This series is known as the exponential series.

exterior angle: when one side of a polygon is extended, the angle between this line and the next side is called the exterior angle. In the diagram below θ is an exterior angle.

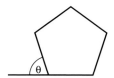

face: (1) a region of the plane bounded by edges of a *planar graph*.

face: (2) one of the plane surfaces of a solid. The diagram shows a tetrahedron. Its faces are the four triangles ABC, ABD, ACD and BCD. A cube has six faces.

D

C

A   B

factors: divide exactly into a number or expression.

For example, the factors of 24 are 1, 2, 3, 4, 6, 8, 12 and 24. The numbers 2 and 3 are prime factors because they are factors of 24 and prime numbers.

An algebraic expression can also have factors. For example, the factors of $x^3 + x$ are x and $x^2 + 1$, because $x^3 + x = x(x^2 + 1)$. The factors of $x^4 + 4x^3 - 7x^2 - 10x$ are x, $x - 2$, $x + 1$ and $x + 5$, because $x^4 + 4x^3 - 7x^2 - 10x = x(x - 2)(x + 1)(x + 5)$. The *factor theorem* can be used to help find the factors of a polynomial. (See also *factorization*.)

factor formulas: the factor formulas are listed below:

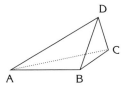

$$\sin A + \sin B = 2 \sin\left(\frac{A + B}{2}\right) \cos\left(\frac{A - B}{2}\right)$$

$$\sin A - \sin B = 2 \sin\left(\frac{A - B}{2}\right) \cos\left(\frac{A + B}{2}\right)$$

$$\cos A + \cos B = 2 \cos\left(\frac{A + B}{2}\right) \cos\left(\frac{A - B}{2}\right)$$

$$\cos A - \cos B = -2 \sin\left(\frac{A + B}{2}\right) \sin\left(\frac{A - B}{2}\right)$$

These formulas are derived from the *trigonometric sum and difference formulas*.

factorization: the process of writing a *polynomial* or other algebraic expression as the product of a number of factors.

If a variable appears in a number of terms, it is a factor and the expression can be factorized. For example in the expression $x^2 + xy$, x is a factor and the expression can be factorized as $x(x + y)$. Similarly $x^3y + xy^2$ can be factorized as $xy(x^2 + y)$, because xy is a factor of both terms.

Quadratic expressions can be factorized as the product of two factors, for example $x^2 + x - 56 = (x - 7)(x + 8)$. Quadratics, such as $x^2 + bx + c$, can be factorized into the form $(x + p)(x + q)$, if it is possible to find p and q such that $pq = c$ and $p + q = b$.

The factors of other polynomials can be found using the *factor theorem* and *division of polynomials*.

Example:

Factorize $x^2 + 2x - 15$.

Solution:

If this is to be factorized into the form $(x + p)(x + q)$, then we need to find p and q so that $pq = -15$ and $p + q = 2$. The first of these suggests that possible values of p and q are ± 1 and ∓ 15 or ± 3 and ∓ 5. However, $p = -3$ and $q = 5$ are the only ones of these that satisfy $p + q = 2$, so $x^2 + 2x - 15 = (x - 3)(x + 5)$.

factor theorem: this states that, for a polynomial $f(x)$, if $f(a) = 0$, then $(x - a)$ is a factor of $f(x)$. This result is useful in helping to find the factors of a polynomial.

Consider the polynomial $f(x) = x^4 - 2x^3 - 5x^2 + 6x$. The values of this polynomial have been tabulated for integer values of x from -3 to 4, and are listed in the table below.

x	-3	-2	-1	0	1	2	3	4
$f(x)$	72	0	-8	0	0	-8	0	72

From the table it can be seen that $f(-2) = 0$, $f(0) = 0$, $f(1) = 0$ and $f(3) = 0$, so the factors are $(x - (-2)) = (x + 2)$, $(x - 0) = x$, $(x - 1)$ and $(x - 3)$. The polynomial can then be written as a product of factors:

$$f(x) = x(x + 2)(x - 1)(x - 3)$$

The graph below shows this polynomial.

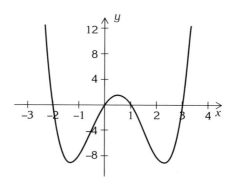

factorial notation: $n!$ is the product of all integers from n to 1 inclusive; that is

$$n! = n(n - 1)(n - 2)(n - 3) \ldots 3 \cdot 2 \cdot 1$$

The symbol $n!$ is read as n factorial. Note that $0!$ is defined as 1.

For example, $6! = 6 \times 5 \times 4 \times 3 \times 2 \times 1 = 720$

feasible flow: a flow through a *network* that satisfies the given capacities on the *edges*.

feasible region: the region within a set of straight-line *graphs* showing linear programming constraints, within which the solution lies.

Fibonacci sequence: the sequence

0, 1, 1, 2, 3, 5, 8, 13, 21, 34, 55, ...

is called the Fibonacci sequence. In the sequence each term is the sum of the two previous terms. The sequence can be defined algebraically as:

$$u_0 = 0, \ u_1 = 1 \quad \text{and} \quad u_{n+1} = u_n + u_{n-1}$$

first-fit algorithm: a sorting *algorithm* for *bin-packing* problems that places an item in the first available bin.

float: the maximum time by which the start of an activity can be delayed without delaying the whole project.

flow augmenting paths: paths in a *network* flow problem that consist entirely of unsaturated arcs.

force: in everyday language we think of a force as something that provides a push or a pull. More precisely, a force is a quantity that produces a change in the motion of an object or attempts to change the motion or equilibrium of an object. A force can cause an object to start moving, stop moving, change speed or change its direction of motion.

Examples of common forces are:

- the *force of gravity* on an object (often called weight)
- the pulling force produced by a string, spring or rod attached to an object, called *tension*
- the pushing force produced by a spring or rod, called the *thrust*
- the contact forces between two touching surfaces; the component of the force at right angles to the surfaces is called the *normal reaction* and the tangential component is called *friction*
- the *air resistance* or *drag force* when an object travels through air
- the *buoyancy force* or *upthrust* on an object placed in a fluid.

In reality an object may experience several forces, so the effect on the object is a combined effect of these forces. The combined force is called the *net* or *resultant force*.

Example 1:

A suitcase is placed on a conveyor belt in the baggage reclaim lounge of an airport. In each of the following cases decide if there is a resultant force acting on the suitcase.

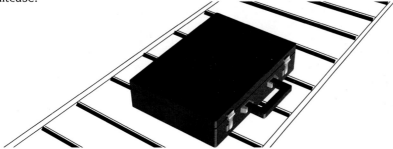

(a) The suitcase is at rest.
(b) The suitcase is moving at constant speed in a straight line.
(c) The suitcase is moving at constant speed around a corner of the conveyor belt.
(d) The suitcase is traveling along a straight part of the conveyor belt and gaining speed.

Solution:

(a) If the suitcase is at rest (then so is the conveyor belt), the resultant force is zero.
(b) If the suitcase is moving at constant speed in a straight line then the resultant force on the suitcase is zero.
(c) If the suitcase is moving at constant speed around a corner of the conveyor belt then its direction of motion is changing so the suitcase experiences a resultant force towards the center of the corner. (This force is produced by the friction between the suitcase and conveyor belt surfaces.)
(d) If the suitcase is traveling along a straight part of the conveyor belt and gaining speed there is a resultant force pointing forwards. (This force is produced by the friction between the suitcase and conveyor belt surfaces.)

Example 2:

Identify the forces on a bungee jumper as she drops from the platform to first coming to instantaneous rest.

Solution:

The figure shows the motion of the bungee jumper split into two phases: (a) when the bungee is slack and (b) when the bungee is stretched.

When the bungee is slack the jumper experiences the force of gravity (downwards) and air resistance (upwards).

When the bungee is stretched the jumper experiences a tension force in the bungee (upwards), the force of gravity (downwards) and air resistance (upwards).

Forces are vector quantities with both a magnitude and a direction and the net or resultant force is found using the rules of vector addition, i.e. the triangle or parallelogram law. In practice it is easier to write a force in terms of components and add the components.

Example 3:

An object P experiences three forces of magnitudes 6 N, 5 N and 9 N as shown in the figure below.

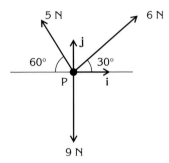

Find the magnitude and direction of the resultant force acting on the object.

Solution:

If unit vectors i and j are chosen horizontally and vertically, respectively, then each force can be written as components in the directions of i and j and the resultant force found by adding the components. The analysis is laid out neatly in the following table.

force	components
6 N	$6 \cos 30° \, i + 6 \cos 60° \, j$
5 N	$-5 \cos 60° \, i + 5 \cos 30° \, j$
9 N	$-9 \, j$
resultant force	$(6 \cos 30° - 5 \cos 60°)i + (6 \cos 60° + 5 \cos 30° - 9)j$

The resultant force acting on the object is $2.70i - 1.67j$ which is shown in the next figure.

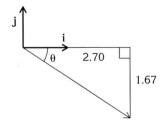

The magnitude of the resultant force is given by $\sqrt{(2.70^2 + 1.67^2)} = 3.17$ N

The direction of the resultant force is defined by the angle θ where

$$\tan(\theta) = \frac{1.67}{2.70} = 0.619$$

giving the direction θ as 32° to the nearest degree.

force of gravity: gravity is the force that the Earth exerts on all objects. If an object has mass m kg and is on or close to the Earth's surface then the force acting on the object has magnitude mg and direction towards the center of the Earth. The term g is a constant known as the acceleration due to gravity. The force that produces this acceleration is called the force of gravity. On Earth g has a value of approximately 9.8 m s^{-2}, sometimes taken as 10 m s^{-2}.

In this model of the force of gravity, it is assumed that the Earth is a sphere and the object is a particle.

The force of gravity acting on an object is often called its weight.

(See also *gravitation*.)

fractions: a fraction is a ratio of two numbers or *polynomials*. The rules for addition, subtraction, multiplication and division of fractions are summarized below:

$$\frac{a}{b} + \frac{c}{d} = \frac{ad + cb}{bd}$$

$$\frac{a}{b} - \frac{c}{d} = \frac{ad - cb}{bd}$$

$$\frac{a}{b} \times \frac{c}{d} = \frac{ac}{bd}$$

$$\frac{a}{b} \div \frac{c}{d} = \frac{a}{b} \times \frac{d}{c} = \frac{ad}{bc}$$

frame of reference: to describe the motion of an object we need to define a *coordinate* system to specify the position of the object and the time. A coordinate system for which the origin and coordinate axes are fixed in space relative to the position of the distant stars is called an *inertial frame of reference*. A set of coordinate axes fixed on the Earth's surface is taken to be a good model for an inertial frame of reference.

free vector: a vector, the position in space of which is not specified. The vector is represented by an arrow (or directed line segment) drawn at any position. The figure below shows several free vectors; they are equal in length and parallel. The vectors are equal. A displacement vector is an example of a free vector.

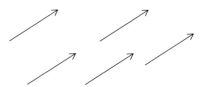

A vector that is not a free vector is called a "bound vector." The position vector of an object is an example of a bound vector because its tail is always drawn from the origin.

frequency curves: a frequency curve is the smooth line graph that is obtained by joining the midpoints of the top of the rectangle that would form a *histogram*.

Example:

Draw a frequency curve of the following data where age is to the nearest year.

Age (years)	0–5	6–10	11–20	21–30	31–50	51–75
Number	35	42	29	45	26	11

Solution:

The *frequency densities* are calculated along with the midpoint of the class:

Class boundaries	0– (5.5)	5.5– (10.5)	10.5– (20.5)	20.5– (30.5)	30.5– (50.5)	50.5– 75.5
Midpoint	2.75	8	15.5	25.5	40.5	63
Frequency density	6.36	8.4	2.9	4.5	1.3	0.44

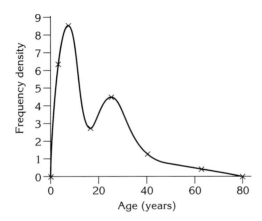

Note that the frequency curve is extended to the neighboring classes having zero frequency.

frequency density: the frequency density of a frequency distribution is given by

$$\frac{\text{class frequency}}{\text{class width}}$$

The frequency density is used when drawing *histograms*, finding the *modal class*, etc. of a frequency distribution. It allows comparisons between classes of unequal width.

Example:

Age (years)	0–5	6–10	11–20	21–30	31–50	51–75
Number	35	42	29	45	26	11

Find the frequency densities of the above data where age is to the nearest year.

Solution:

Class	0–	5.5–	10.5–	20.5–	30.5–	50.5–
boundaries	(5.5)	(10.5)	(20.5)	(30.5)	(50.5)	75.5
Class widths	5.5	5	10	10	20	25
Frequency density	35/5.5 = 6.36	42/5 = 8.4	2.9	4.5	1.3	0.44

frequency distribution: a frequency distribution is created when observations are made on a variable. The variable may be *discrete* or *continuous*. Frequency distributions are recorded in a *frequency table*.

frequency polygon: a frequency polygon is the line graph that is obtained by joining the midpoints of the top of the rectangle which would form a *histogram*. It is created in exactly the same way as the *frequency curve* but the points are joined by straight lines.

Below is the frequency polygon for the *frequency curves* example.

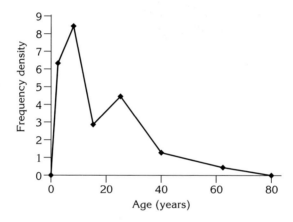

frequency table: a *frequency distribution* is recorded in a frequency table. The frequency table gives the possible values or class intervals of the variable and the corresponding frequency.

The following is a frequency table of the test scores of a class of 50 students.

Score	0–9	10–19	20–29	30–39	40–49	50–59	60–69	70–79	80–89	90–100
f	0	2	2	5	8	10	12	6	4	1

friction: imagine that you are trying to push a very heavy suitcase across the floor and suppose that it will not move. There must be a horizontal force acting on the suitcase to balance your pushing force. This force is called the "force of friction."

Two surfaces in contact experience a normal reaction force R and a tangential force F called the force of friction.

The force of friction opposes the direction of relative motion between the surfaces.

For a static particle, the force of friction is just sufficient to stop motion; the law of friction is $F \leq \mu_s R$ where μ_s is the coefficient of static friction, formulated by *Coulomb*.

For a moving particle, experimental evidence suggests the force of friction is directly proportional to the normal reaction force, $F = \mu_d R$ where μ_d is the coefficient of dynamic friction.

There is experimental evidence that $\mu_d < \mu_s$.

Example:

The coefficient of friction between a suitcase of mass 50 kg and the floor is 0.6. What is the least horizontal force that you need to apply to push the suitcase across the floor?

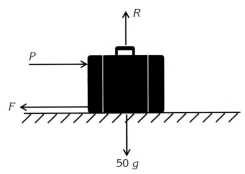

$$50\,g$$

Solution:

The forces acting on the suitcase when it is sliding across the floor are the force of gravity, normal reaction, R, friction, F, and the horizontal push, P. These are shown in the figure above.

If the suitcase slides across the floor then:

$R = 50\,g$ and $P > F$

Since the suitcase is moving the model of dynamical friction applies,

$F = 0.6\,R = 0.6 \times 50\,g = 30\,g$

Thus $P > 30\,g$

The least horizontal force that will move the suitcase is $30\,g = 294$ newtons.

frustrum: a frustrum is formed when a solid such as a cone or pyramid is truncated to leave a solid with two parallel faces. The diagram below shows a frustrum formed by truncating a cone.

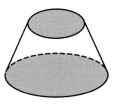

full-bin algorithm: a sorting *algorithm* for bin-packing problems that fills bins one at a time.

function: a function is a relationship between two sets that associates each element of the first set with only one element of the second set. If x is an element of the first set, then $f(x)$ is used to refer to the associated element of the second set, and is called the "image of x." The first set is called the domain of the function and the second set is called the codomain of the function. A subset of the codomain which is actually used by the function is called the *range* of the function. For example, consider the domain {1, 2, 3, 4}, the codomain {1, 4, 5, 9, 10, 11, 12, 16, 20} and the rule $f(x) = x^2$. Then the range of the function $f(x) = x^2$ is the set {1, 4, 9, 16}.

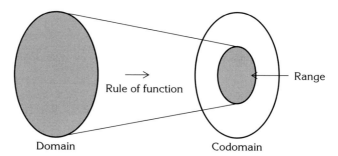

A function can be one to one or many to one.

For example the function $f(x) = x^2$ is a many to one function. The diagram below illustrates this function for a specific domain.

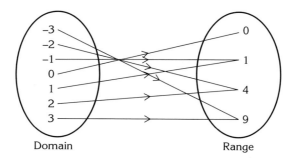

The function $f(x) = x + 1$ is a one to one function. It is illustrated below for a specific domain.

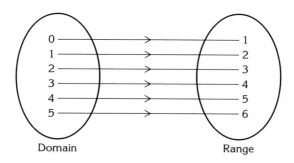

Gaussian elimination method: a method of solving simultaneous equations by eliminating one variable and solving for the other – it can also be used for systems of more than two equations, using a similar principle. The method is named after the famous mathematician Carl Friedrich Gauss (1777–1855).

Example:

Solve the equations:

$$x + 2y = 7$$
$$2x + 7y = 23$$

Solution:

Elimination stage:
$$\begin{bmatrix} 1 & 2 & 7 \\ 2 & 7 & 23 \end{bmatrix} \begin{matrix} R_1 \\ R_2 \end{matrix} \implies \begin{matrix} R_1 \\ R_2 - 2R_1 \end{matrix} \begin{bmatrix} 1 & 2 & 7 \\ 0 & 3 & 9 \end{bmatrix}$$

Back substitution stage: $\implies y = 3; \quad x = 1$

general iterative method: to solve the equation $f(x) = 0$, rearrange it in the form $x = g(x)$ and then use the iterative scheme $x_{n+1} = g(x_n)$. This will generate a sequence that may converge to the solution of the required equation. The sequence does not converge if $|g(x)| > 1$ near to the root.

Example:

The equation $x^3 - 9x + 4 = 0$ has a solution close to $x = 0$. Starting with $x_0 = 0$, use an iterative scheme of the form

$$x_{n+1} = \frac{a + x^3_n}{b}$$

to find this root correct to three decimal places.

Solution:

Start by rearranging $x^3 - 9x + 4 = 0$. This gives:

$$x^3 - 9x + 4 = 0$$
$$9x = 4 + x^3$$
$$x = \frac{4 + x^3}{9}$$

Using the scheme $x_{n+1} = \frac{4 + x^3_n}{9}$ and $x_0 = 0$ gives:

$$x_1 = \frac{4 + 0}{9}$$

$$= 0.4444$$

Iterating with a calculator then gives:

$$x_2 = 0.45420$$
$$x_3 = 0.45486$$
$$x_4 = 0.45490$$

So the solution is $x = 0.455$ (to 3 decimal places)

Note that other possible rearrangements are possible, but that these may or may not converge or may converge to a different root.

general solution of a differential equation: this will contain arbitrary constants. For example $y = Ae^{0.2t}$ is the general solution of the differential equation

$$\frac{dy}{dt} = 0.2y$$

The general solution describes a family of curves, with each curve corresponding to a particular value of A. The diagram shows the family of curves for the differential equation above.

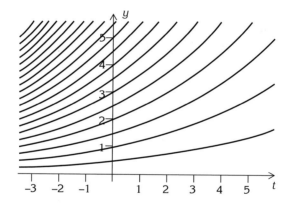

geometric mean: a measure of location or average. The geometric mean of the n numbers $x_1, x_2, \ldots x_n$ is given by:

$$\sqrt[n]{(x_1 \times x_2 \times \ldots \times x_n)}$$

Example:

Find the geometric mean for the following set of numbers:

4, 6, 2, 3, 5, 7, 1, 8, 3, 6, 2, 4.

Solution:

The geometric mean is given by

$$\sqrt[12]{(4 \times 6 \times 2 \times 3 \times 5 \times 7 \times 1 \times 8 \times 3 \times 6 \times 2 \times 4)}$$

$$= \sqrt[12]{(5806080)}$$

$$= 3.6615$$

For n values x_1, x_2, ... x_n with frequency f_1, f_2, ... f_n the geometric mean is given by

$$\Sigma f\sqrt{(x_1^{f_1} \times x_2^{f_2} \times \ldots \times x_n^{f_n})}$$

geometric progression (GP): this is a sequence or series of terms or numbers, where each term is obtained by multiplying the previous term by the same constant, known as the common ratio. An example of a geometric sequence and a geometric series are given below.

    1, 2, 4 , 8, 16, 32, 64, 128, ...

    2 + 2.2 + 2.42 + 2.662 + 2.9282 + ...

The first term of a GP is usually denoted by the letter a and the common ratio as r. The nth term of a GP can be found using the formula:

$$u_n = ar^{n-1}$$

The sum of the first n terms, S_n, of a GP can be found using the formula:

$$S_n = \frac{a(1 - r^n)}{1 - r}$$

The sum of an infinite GP will converge if $r < 1$. In this case the sum of the infinite GP is given by

$$S_\infty = \frac{a}{1 - r}$$

Example:

For the geometric sequence,
    100, 90, 81, 72.9, 65.61, ...
find:
(a) the 10th term
(b) the sum of the first 20 terms
(c) the sum to infinity.

Solution:

First note that, in this GP, $a = 100$ and $r = 0.9$.

(a) Using $u_n = ar^{n-1}$ with $n = 10$, $a = 100$ and $r = 0.9$ gives:

$$u_{10} = 100 \times 0.9^9$$
$$= 38.742 \text{ to 3 decimal places}$$

(b) Using,

$$S_n = \frac{a(1 - r^n)}{1 - r} \text{ with } n = 20, a = 100 \text{ and } r = 0.9 \text{ gives:}$$

$$S_{20} = \frac{100(1 - 0.9^{20})}{1 - 0.9}$$
$$= 878.423 \text{ to 3 decimal places}$$

(c) Using

$$S_\infty = \frac{a}{1 - r}$$

with $a = 100$ and $r = 0.9$ gives:

$$S_\infty = \frac{100}{1 - 0.9}$$

$$= 1000$$

geometric series: see *geometric progression*.

geometry: the study of properties of shapes, for example circles, triangles and squares, and is used to reach conclusions about the sizes of angles and the lengths of lines. Geometry is not restricted to shapes in two dimensions, and many results used in two dimensions can be extended to three dimensional situations.

goodness of fit: the chi squared (χ^2) test is used to test the goodness of fit of a set of observed frequencies to a hypothesized distribution, i.e. *binomial, Poisson* or *normal*.

In order to use the test the total frequency must be "large" (>50), and the individual class frequencies must be greater than 5. Adjoining classes should be combined to achieve this frequency.

The observed frequencies are calculated under the *null hypothesis* H_0, the χ^2 test performed and the resulting *test statistic* examined. The *degrees of freedom* v, are found from

v = number of classes (n) – number of parameter restrictions – 1

The restrictions are the number of parameters needed to calculate the expected frequencies which had to be estimated from the observed frequencies. These are summarized below for the *binomial, Poisson* and *normal distributions*.

	Parameter restrictions	Degrees of freedom v
Binomial:	p known	$n - 1$
	p estimated from observed frequencies	$n - 2$
Poisson:	λ known	$n - 1$
	λ estimated from observed frequencies	$n - 2$
Normal:	μ and σ known	$n - 1$
	μ and σ estimated from observed frequencies	$n - 3$

Example 1:

In a seed viability trial to test the seed producers' claim that the germination of the seeds was binomially distributed with a probability of germination of 0.5, 600 seeds were planted in 100 rows of 6. The number of seeds that germinated in each row was counted.

Number of seeds germinating	0	1	2	3	4	5	6
Number of rows	1	4	7	29	33	18	8

Test the producers' claim.

Solution:

$H_0 : p = 0.5, \quad H_1 : p \neq 0.5$

The number of seeds germinating in each row, X, is a binomial situation with $n = 6$ and $p = 0.5$. Therefore

$$P(X = x) = \binom{n}{x} p^x (1 - p)^{n-x}, \; x = 1, 2, 3, 4, 5, 6$$

The expected frequencies are calculated using the binomial formula above.

x	$P(X = x)$	Expected frequency $P(X = x) \times 100$
0	0.0156	1.56
1	0.0938	9.38
2	0.2344	23.44
3	0.3125	31.25
4	0.2344	23.44
5	0.0938	9.38
6	0.0156	1.56

The categories for $x = 0$ and $x = 1$ must be combined as the expected frequency for $x = 0$ is less than 5. Similarly with categories for $x = 5$ and $x = 6$.

The χ^2 test statistic, $\chi^2 = \sum\limits_{i=1}^{n} \dfrac{(O_i - E_i)^2}{E_i}$ is calculated:

x	Observed (O)	Expected (E)	$\dfrac{(O - E)^2}{E}$
1 and less	5	10.94	3.225
2	7	23.44	11.53
3	29	31.25	0.162
4	33	23.44	3.899
5 or more	26	10.94	20.732
			39.548

Here the parameter p was known and so the degrees of freedom $v = 7 - 1 = 6$.

The 5% critical value with $v = 6$ is 12.592. The calculated value is more than the critical value and so this leads to the rejection of H_0 in favor of H_1. Therefore it can be concluded that the claim that germination was binomially distributed with probability of germination being 0.5 is false.

Example 2:

The quality control department of a manufacturing firm randomly selects 200 components and tests them for defects.

Number of defects	0	1	2	3	4	>5
Number of components	90	75	31	3	1	0

If the process is working properly the number of defects should be distributed with a Poisson distribution. Test this hypothesis.

Solution:

H_0 : the data fits a Poisson distribution
H_1 : the data does not fit a Poisson distribution.

The parameter for the Poisson distribution is λ, the mean number of defects. This has to be estimated from the data. The mean of a frequency distribution is given by

$$\lambda = \frac{\Sigma fx}{n} = \frac{\Sigma(90 \times 0 + 75 \times 1 + 31 \times 2 + 3 \times 3 + 1 \times 4 + 0 \times 5)}{200} = \frac{150}{200} = 0.75$$

The probability of 0, 1, 2, 3 and 4 defects are calculated using the Poisson formula. The probability of 5 or more defects are calculated from $1 - P(4 \text{ or less})$.

$$P(X = x) = \frac{e^{-\lambda}\lambda^x}{x!}$$

Number of defects x	$P(X = x)$	Expected frequency $P(X = x) \times 200$
0	0.4724	94.48
1	0.3543	70.86
2	0.1329	26.58
3	0.0332	6.640
4	0.0062	1.240
5 or more	0.0010	0.200

The χ^2 test statistic, $\chi^2 = \sum_{i=1}^{n} \frac{(O_i - E_i)^2}{E_i}$

is calculated, but first the last three classes must be grouped together to ensure that the expected frequencies are greater than 5:

Number of defects	Observed (O)	Expected (E)	$\frac{(O - E)^2}{E}$
0	90	94.48	0.2124
1	75	70.86	0.2419
2	31	26.58	0.7350
3 and over	4	8.08	2.0602
			3.2495

Here the parameter λ was estimated from the observed frequencies and so the degrees of freedom $v = 4 - 2 = 2$.

The 5% critical value with $v = 2$ is 5.991. The calculated value is less than the critical value and so this leads to the acceptance of H_0. Therefore it can be concluded that the number of defects follows a Poisson distribution.

googol: the name given to the number that consists of a 1 followed by 100 zeros, that is 10^{100}.

googolplex: the name given to the number that consists of a 1 followed by a googol of zeros, that is $10^{10^{100}}$.

gradient: The gradient of a straight line with equation $y = mx + c$ is m. (See also *differentiation*.)

graph: a diagram of *vertices* and *edges* representing how objects are related to each other.

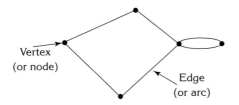

Vertex
(or node)

Edge
(or arc)

gravitation: Isaac Newton (1643–1727) formulated the law of universal gravitation between two objects. The law states that between two objects of masses m_1 and m_2, with centers of mass a distance d apart, there is an attractive force of magnitude:

$$F = \frac{Gm_1m_2}{d^2}$$

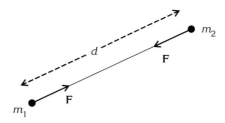

G is the gravitational constant and in SI units has a value 6.67×10^{-11} kg^{-1} m^3 s^{-1}. The force **F** is called the gravitational force.

The gravitational force on an object on (or close to) the surface of the Earth is often called the weight of the object.

Example 1:

Calculate the magnitude of the gravitational force between the following objects:

(a) a boy of mass 63 kg and a girl of mass 50 kg who are 1 meter apart
(b) the Moon (mass 7.38×10^{22} kg) and the Earth (mass 5.98×10^{24} kg) if the distance between their centers is 3.84×10^8 m.

Solution:

(a) $F = \dfrac{6.67 \times 10^{-11} \times 63 \times 50}{1^2} = 2.1 \times 10^{-7}$ newtons

(b) $F = \dfrac{6.67 \times 10^{-11} \times 5.98 \times 10^{24} \times 7.38 \times 10^{22}}{(3.84 \times 10^8)^2} = 2.0 \times 10^{20}$ newtons

Example 2:

Find the magnitude of the gravitational force on an object of mass M on the Earth's surface. Assume that the Earth is a sphere of mass 5.98×10^{24} kg and radius 6.37×10^6 m.

Solution:

The magnitude of the force is

$$F = \frac{6.67 \times 10^{-11} \times 5.98 \times 10^{24} \times M}{(6.37 \times 10^6)^2} = 9.8M \text{ newtons}$$

We see that Newton's universal law of gravitation gives the familiar rule for the force of gravity or the weight of an object of mass M as Mg, where $g = 9.8$ m s^{-2}.

gravity: the word gravity often has two meanings:

- The acceleration of an object close to the Earth's surface is commonly referred to as the acceleration due to gravity and is denoted by g. Its value is approximately 9.8 m s^{-2} and for ease of calculation is taken as 10 m s^{-2} in many problems.

- The force of gravity on a object of mass M close to the Earth's surface which is equal to $9.8M$ newtons.

More generally we talk about gravity on any planet. This refers to either the acceleration or the force on an object close to that planet. For example, the gravity on the Moon would refer to either the acceleration of an object close to the Moon's surface (1.6 m s^{-2}) or the force on an object of $1.6M$ newtons.

great circle: a great circle is a circle drawn on the surface of a sphere, so that it has the same radius as the sphere.

greatest common divisor (GCD): the greatest common divisor of two numbers is the largest number that will divide both numbers. The GCD is also known as the *greatest common factor* (GCF) or *highest common factor* (HCF). See *highest common factor* for more detail.

greatest common factor: See *highest common factor*.

greedy algorithm: an *algorithm* that chooses the best option at each stage. Once an item is included in the solution, it will not be changed.

growth: many situations in which growth is considered can be modeled using *exponential growth*.

half-angle formula: these formulas are trigonometric identities used to give sin A, cos A and tan A in terms of tan $\frac{1}{2} A$

$$\sin A = \frac{2\tan \frac{1}{2}A}{1 + \tan^2 \frac{1}{2}A}$$

$$\cos A \ = \frac{1 - \tan^2 \frac{1}{2}A}{1 + \tan^2 \frac{1}{2}A}$$

$$\tan A = \frac{2\tan \frac{1}{2}A}{1 - \tan^2 \frac{1}{2}A}$$

Hamiltonian cycle: a cycle that passes through every *vertex* of a *graph* once, and only once, and returns to the starting point.

handshaking lemma: the sum of the degrees of the *vertices* in a *graph* equals twice the number of *edges* in the *graph*.

harmonic mean: the harmonic mean of the n values x_1, x_2, ... x_n is given by

$$\frac{n}{\sum \frac{1}{x}}$$

For n values with respective frequencies f_1, f_2, ... f_n the harmonic mean is given by

$$\frac{\sum f}{\sum \frac{f}{x}}$$

Example:

Find the harmonic mean of the numbers 3, 5, 4, 2, 5, 6, 4, 3.

Solution:

$$\text{harmonic mean} \ = \frac{8}{\frac{1}{3} + \frac{1}{5} + \frac{1}{4} + \frac{1}{2} + \frac{1}{5} + \frac{1}{6} + \frac{1}{4} + \frac{1}{3}} = 3.582$$

helix: a helix or spiral is a curve that lies on the surface of a cylinder or cone. The parametric equations of a circular helix, that lies on the surface of a cylinder of radius a, are $x = a \cos t$, $y = a \sin t$ and $z = bt$. See the diagram on page 108.

hemisphere: a hemisphere is formed when a sphere is cut in half by a plane that passes through the center of the sphere.

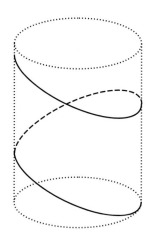

Helix

Heron's formula: (also known as Hero's formula) used to find the area of a triangle in which the lengths of all the sides are known. If the lengths of the sides are a, b and c, then Heron's formula states that the area is given by:

$$A = \sqrt{s(s-a)(s-b)(s-c)}$$

where

$$s = \tfrac{1}{2}(a + b + c)$$

heuristic algorithm: an *algorithm* that produces a good solution to a problem using logical and intuitive steps.

highest common factor (HCF): of two numbers is the largest number that will divide into both numbers. For example, the highest common factor of 35 and 50 is 5. To find the highest common factor it is useful to compare the numbers when they are written as the product of primes. For example, to find the HCF of 882 and 3006, note that $882 = 2 \times 3^2 \times 7^2$ and $3006 = 2 \times 3^2 \times 167$. Then the HCF is $2 \times 3^2 = 18$, using the prime factors that are common to both numbers.

histogram: a histogram is a pictorial representation of a *continuous frequency distribution*.

It consists of rectangles on a continuous base line. The area of each rectangle is proportional to the frequency of the class it represents. The extremes of the base of each rectangle are the lower *class boundary* and the upper class boundary of the class that it represents. The height of each rectangle is found from the frequency density:

$$\frac{\text{class frequency}}{\text{class width}}$$

Example:

Draw a histogram for the following data of lengths of rods given to the nearest meter.

Length (meters)	0–5	6–10	11–20	21–30	31–50	51–75
Number	35	42	29	45	26	11

Solution:

Class boundaries	0– (5.5)	5.5– (10.5)	10.5– (20.5)	20.5– (30.5)	30.5– (50.5)	50.5– 75.5
Class widths	5.5	5	10	10	20	25
Frequency density	6.36	8.4	2.9	4.5	1.3	0.44

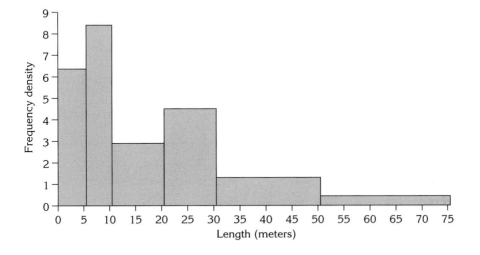

Hooke's law: an experimental law relating the tension in an elastic string or spring to its extension.

If a string or spring is stretched so that its extension is x meters and it exerts a tension T on an object attached to the end of the string (or spring) then Hooke's law states that T is directly proportional to x

$$T = kx$$

The constant of proportionality, k, is called the stiffness of the string or spring. The units of stiffness are newtons per meter (N m^{-1}). Sometimes the stiffness is written as λ/l where λ is called the *modulus of elasticity* and l is the natural length of the string (or spring).

An elastic string or spring that satisfies Hooke's Law is called a *perfect string* or *perfect spring*. This reminds us that Hooke's law is only a model for real strings and springs. It assumes that a string or spring returns to its natural length when the tension is removed.

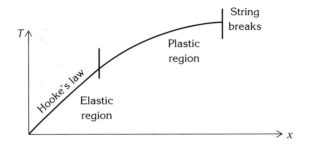

Hooke's law usually breaks down when the extension is too large; what happens is that the string (or spring) loses its elasticity and does not return to its natural length.

The figure on the previous page shows the typical experimental results for an elastic string stretched until it snaps.

hyperbola: a hyperbola has the equation

$$\frac{x^2}{a^2} - \frac{y^2}{b^2} = 1$$

Its shape is shown in the graph below.

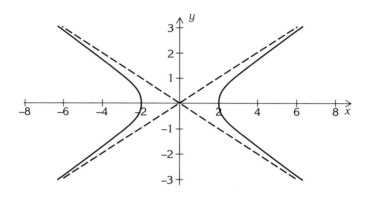

Note the two *asymptotes* that have been added to the diagram. These have equations

$$y = \frac{bx}{a} \quad \text{and} \quad y = -\frac{bx}{a}$$

In parametric form the equations of the hyperbola are $x = a \sec t$ and $y = b \tan t$

hypotenuse: in a right angled triangle the hypotenuse is the side opposite the right angle. In the diagram BC is the hypotenuse.

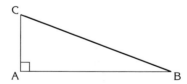

hypothesis: a statistical hypothesis is an assumption about the value of a *parameter* of the distribution. There are two types of hypotheses: the *null hypothesis* and the *alternative hypothesis*.

For example, consider a sample of peas from a species which was known to have a mean mass of 0.1 g. We wish to test if the mean mass of peas in the sample differs from 0.1 g.

Here the *null hypothesis* is $H_0:\mu = 0.1$ and the *alternative hypothesis* is $H_1:\mu \neq 0.1$.

hypothesis testing: a hypothesis test or *significance test* is a rule which decides on the acceptance or rejection of the *null hypothesis* H_0, based on the results of a *random sample* of the population under consideration. The *chi squared test* and the *Student t test* are examples of hypothesis tests.

identity: an identity is a result that is true for all values of the variables involved, unlike an equation which is true only for specific values of the variables. Examples of identities are:

$$\cos^2 \theta + \sin^2 \theta \equiv 1 \qquad \text{and} \qquad \frac{1}{x} + \frac{1}{y} \equiv \frac{x + y}{xy}$$

In an identity the sign $\equiv$ can be used instead of the sign $=$, as shown above.

The statement $\cos^2 \theta + \sin^2 \theta \equiv 1$ is an identity because it is true for all values of θ. The equation $2 \sin \theta = 1$ is not an identity, because it is only true for certain values of θ, such as 30° and 150°.

iff: an abbreviation for if and only if.

See also *necessary and sufficient*.

image: the image of a *function* is the value of the function for a certain value of the *domain*. For example, consider the function $f(x) = x^2$. In this case the image of 2 is 4.

imaginary number: this is any number of the form ai where a is a real number other than 0 and $i = \sqrt{-1}$

imaginary part: The imaginary part of a *complex number* is the coefficient of i. For the complex number $z = a + bi$, the imaginary part is b. The notation $\text{Im}(z) = b$ is often used.

implicit differentiation: an equation like $x^2 + y^2 = 9$ is called an implicit equation. Equations like this can be differentiated with respect to x on a term by term basis. When differentiating note that:

y differentiates to $\dfrac{dy}{dx}$

y^n differentiates to $ny^{n-1}\dfrac{dy}{dx}$

and, in general,

$f(y)$ differentiates to $f'(y)\dfrac{dy}{dx}$

The *product rule* can be used to differentiate expressions like x^2y.

$$\frac{d}{dx}(x^2y) = x^2 \times \frac{dy}{dx} + 2x \times y$$

$$= x^2 \frac{dy}{dx} + 2xy$$

This process is known as implicit differentiation and dy/dx can often be made the subject of the resulting expression.

Example:

Find $\dfrac{dy}{dx}$ for $x^2 + y^2 = 9$.

Solution:

Differentiating $x^2 + y^2 = 9$ with respect to x gives

$$2x + 2y\dfrac{dy}{dx} = 0$$

This can then be rearranged to make $\dfrac{dy}{dx}$ the subject,

$$2x + 2y\dfrac{dy}{dx} = 0$$

$$2y\dfrac{dy}{dx} = -2x$$

$$\dfrac{dy}{dx} = -\dfrac{x}{y}$$

improper fraction: a fraction for which the term in the numerator is greater than the term in the denominator.

If an improper fraction is made up of polynomials

$$\dfrac{P(x)}{Q(x)}$$

then the degree of $P(x)$ is greater than the degree of $Q(x)$. For example,

$$\dfrac{2x^9 + 7x^2 + 5x^2 + 12}{6x^5 + 4x - 9}$$

is an improper fraction as degree of numerator (= 9) > degree of denominator (= 5).

impulse: if a particle of mass m moving with initial velocity $\mathbf{v}_B$ experiences a force $\mathbf{F}$ for a short time Δt which produces a final velocity $\mathbf{v}_A$ then the impulse of the force $\mathbf{F}$ is defined by the change in linear momentum of the particle

impulse $\mathbf{I} = m\mathbf{v}_A - m\mathbf{v}_B$

(momentum after impulse – momentum before impulse).

Impulse is a vector quantity and has SI units kg m s^{-1} or N s.

Example:

A car of mass 1 tonne is involved in a collision in which it is brought to rest from a speed of 10 m s^{-1}. Find the impulse on the car.

Solution:

The momentum before the collision is $1000 \times 10 = 10\,000$ kg m s^{-1}

The momentum after the collision is zero (the car is brought to rest).

Impulse $= 0 - 10\,000$

The magnitude of the impulse is 10 000 kg m s^{-1} and the direction is opposed to the initial velocity of the car. (This makes senses since to stop the car we would need a force opposed to its direction of travel.)

To relate the impulse to the actual force acting, consider Newton's second law (see *Newton's laws of motion*) during the duration of the force $0 \leq t \leq \Delta t$:

$$\mathbf{F} = m\frac{d\mathbf{v}}{dt} = \frac{d}{dt}(m\mathbf{v})$$

Integrating both sides with respect to t between 0 and Δt, we have

$$\int_0^{\Delta t} \mathbf{F}\, dt = \int_{v_B}^{v_A} m\mathbf{v}\, dv = m\mathbf{v}_A - m\mathbf{v}_B$$

An alternative definition of the impulse of a force $\mathbf{F}$ acting for a time Δt is the integral

$$\mathbf{I} = \int_0^{\Delta t} \mathbf{F}\, dt$$

For a constant force this simplifies to $\mathbf{I} = \mathbf{F}\,\Delta t$.

Example:

A car of mass 1 tonne is involved in a collision in which it is brought to rest from a speed of 10 ms^{-1}. Find the average force acting on the car if it takes 1.5 seconds to stop.

Solution:

From the previous solution the magnitude of the impulse acting on the car is

10 000 kg m s^{-1}

Suppose that the average force that leads to this impulse has magnitide F, then

$$F \times 1.5 = 10\ 000$$

Hence $F = 6667$ newtons.

increasing function: a function is said to be an increasing function on an interval if, for any values a and b in the interval where $a < b$ we have $f(a) < f(b)$. Sometimes a function is described as being increasing in an interval, (a, b) if $f'(x) > 0$ for $a < x < b$. For example, $f(x) = x^2$ is an increasing function in the interval $0 < x < \infty$ because $f'(x) = 2x$, which satisfies $f'(x) > 0$ for $0 < x < \infty$.

indefinite integral: if a function $F(x)$ is such that its derivative with respect to x is $f(x)$, then the indefinite integral of $f(x)$ is $F(x) + c$. This can be expressed as:

$$\int f(x)\, dx = F(x) + c$$

Note that the indefinite integral does not involve limits of integration and contains a constant c known as the constant of integration.

Example:

Find the following indefinite integrals:

(a) $\int x^2 + 6 \, dx$

(b) $\int \cos(2x) \, dx$

Solution:

(a) $\int x^2 + 6 \, dx = \dfrac{x^3}{3} + 6x + c$

(b) $\int \cos(2x) \, dx = \dfrac{1}{2}\sin(2x) + c$

independence test: the independence of two variables can be tested using a *contingency table* with a *chi squared* test.

independent events: two *events* are said to be independent if the occurrence of one has no effect on the occurrence of the other. For example, a die is thrown and the score noted. The die is then thrown again. The result of the first score has no effect on the second score and so the two events (the scores) are independent.

independent float: when the activities with *floats* in a *precedence network* are independent of each other.

index numbers: a set of numbers can be reduced to relative values by comparing each number in the set to a fixed base number. It is these relative numbers that are called index numbers or *percentage relative numbers*. If they refer to prices they are sometimes called *price relatives*.

An index number is calculated using $q_n / q_0 \times 100$, where q_0 is the quantity in the base time period (year, month, etc.) and q_n is the quantity in the other time period.

Example:

A family's food costs over a three year period are shown below.

Year 1	Year 2	Year 3
2000	2350	2500

Calculate the index number of each year, with year 1 as a base.

Solution:

For year 2 the index number is $\dfrac{2350}{2000} \times 100 = 117.5$

For year 3 the index number is $\dfrac{2500}{2000} \times 100 = 125$

indices: in the statement $x \times x \times x \times x \times x = x^5$, the number 5 is the index. In general, in a term like x^n, n is called the index. The terms "power" or "exponent" are sometimes used instead of index. The term "indices" is used when there is more than one index.

The following rules are important when dealing with indices:

$$a^m \times a^n = a^{m+n}$$

$$a^m \div a^n = \frac{a^m}{a^n} = a^{m-n}$$

$$(a^m)^n = a^{m \times n}$$

$$a^{-n} = \frac{1}{a^n}$$

$$a^{1/n} = \sqrt[n]{a}$$

Example:

Simplify

(a) $5^3 \times 5^5$

(b) $\dfrac{3^8}{3^6}$

(c) $(2^4)^3$

(d) 3^{-2}

(e) $32^{1/5}$

Solution:

(a) $5^3 \times 5^5 = 5^{3+5}$
$= 5^8$
$= 390625$

(b) $\dfrac{3^8}{3^6} = 3^{8-6}$
$= 3^2$
$= 9$

(c) $(2^4)^3 = 2^{4 \times 3}$
$= 2^{12}$
$= 4096$

(d) $3^{-2} = \dfrac{1}{3^2}$
$= \dfrac{1}{9}$

(e) $32^{1/5} = \sqrt[5]{32}$
$= 2$

induction: see *proof by induction*.

inelastic collision: when two objects collide they either separate and remain as two objects or they coalesce and become one object. If they separate then the collision is called an *elastic collision*; otherwise it is an *inelastic collision*.

inequalities: an inequality is an expression that includes one of the following symbols:

> greater than

≥ greater than or equal to

< less than

≤ less than or equal to

Inequalities, in one variable, can be represented on a number line, as shown in the examples below:

$x \geq -2$

$x < 3$

Note the different symbols used when the end point is included and when the end point is not included.

Inequalities in two variables can be represented on a graph, as a region. The graph below shows the region $x + y \geq 3$.

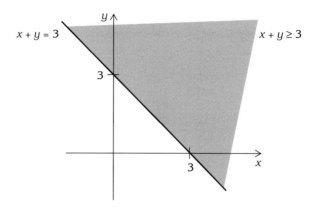

inertia: the inertia of a body is its reluctance to undergo a linear acceleration.

infinite face: the area outside the *edges* of a *planar graph*.

infinite sequence: this is a sequence of numbers that contains an infinite number of elements. For example, the sequence 2, 4, 8, 16, 32, 64, ... is an infinite sequence. Note that this sequence can be specified in three different ways:

- a partial list 2, 4, 8, 16, 16, 32, ...
- a recurrence relation $u_{n+1} = 2u_n$, with $u_1 = 2$
- a function $u(n) = 2^n$ for n a positive integer.

initial conditions: provide information about the state of a system at time $t = 0$. These arise very often in mechanics problems. The initial conditions can be used to find the constants that arise as a result of integrating an expression or solving a differential equation. An example of a set of initial conditions is $x = 5$ and $v = 2$, when $t = 0$.

inscribed circle: an inscribed circle is a circle drawn inside a triangle, so that each side of the triangle is a tangent to the circle. The center of the inscribed circle is at the point where the angle bisectors of the triangle intersect.

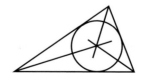

integer: an integer is a number that can be expressed as the sum or difference of two *natural numbers*. Some examples of integers are –5, –4, –3, –2, –1, 0, 1, 2, 3, 4, 5, … .

The set of all integers includes all the natural numbers.

integer part: the integer part of a number is the largest integer that is less than or equal to that number. For example, the integer part of 8.71 is 8 and the integer part of –5.82 is –6.

integration: the rule for integration is the reverse process to differentiation. For example, you are given the derivative and need to find the function. If the derivative of an unknown function with respect to x is $f(x)$, then the function is $F(x)$, where

$$\int f(x)dx = F(x) + c$$

where c is the constant of integration.

The following table shows the standard integrals

Function	Integral
x^n	$\dfrac{x^{n+1}}{n+1}$ $n \neq -1$
$\dfrac{1}{x}$	$\ln|x|$
$\cos x$	$\sin x$
$\sin x$	$-\cos x$
e^x	e^x

Example:

Integrate $\displaystyle\int\left(e^{3x} + \frac{1}{x^4}\right)dx$

Solution:

$$\int\left(e^{3x} + \frac{1}{x^4}\right)dx = \int e^{3x}dx + \int x^{-4}dx + c$$

$$= \tfrac{1}{3}e^{3x} - \tfrac{1}{3}x^{-3} + c$$

$$= \tfrac{1}{3}e^{3x} - \frac{1}{3x^3} + c$$

integration by direct substitution: some integrals can be reduced to a standard form by using a substitution. An integral with respect to x can be changed to an integral with respect to u by using the following result:

$$\int f(x)dx = \int F(u)\frac{dx}{du}du$$

The application of this approach is shown in the following example.

117

integration by direct substitution

Example 1:

Find $\int \sin^4 x \cos x \, dx$

Solution:

Use the substitution $u = \sin x$, which gives

$$\frac{du}{dx} = \cos x \qquad \text{and} \qquad \frac{dx}{du} = \frac{1}{\cos x}$$

With this substitution the integral becomes:

$$\int \sin^4 x \cos x \, dx = \int u^4 \cos x \, \frac{1}{\cos x} \, du$$

$$= \int u^4 \, du$$

$$= \frac{u^5}{5} + c$$

$$= \tfrac{1}{5} \sin^5 x + c$$

Note that the final result is given in terms of x, rather than leaving it in terms of u.

The following example shows how to deal with the limits of integration when finding a definite integral with this method.

Example 2:

Find $\int_2^3 x e^{x^2} \, dx$

Solution:

Use the substitution $u = x^2$. Then $\frac{du}{dx} = 2x$ and $\frac{dx}{du} = \frac{1}{2x}$.

Before making the actual substitution, note that,

when $x = 2$, $u = 2^2 = 4$ and, when $x = 3$, $u = 3^2 = 9$.

Then, using the substitution gives:

$$\int_2^3 x e^{x^2} \, dx = \int_4^9 x e^u \, \frac{1}{2x} \, du$$

$$= \int_4^9 \frac{e^u}{2} \, du$$

$$= \left[\frac{e^u}{2} \right]_4^9$$

$$= \frac{e^9}{2} - \frac{e^4}{2}$$

integration of e^x: the integral of e^x is simply $\int e^x \, dx = e^x + c$.

Further, $\int e^{kx} \, dx = \dfrac{1}{k} e^{kx} + c$

Example:

Find the following:

(a) $\int e^{2x} \, dx$

(b) $\int_0^1 4e^{x/2} \, dx$

Solution:

(a) $\int e^{2x} \, dx = \dfrac{1}{2} e^{2x} + c$

(b) $\int_0^1 4e^{x/2} \, dx = \left[4 \times 2e^{x/2} \right]_0^1$

$\qquad\qquad\qquad = 8e^{1/2} - 8 = 5.19$

integration by parts: some products of functions can be integrated by using the method of integration by parts. This method produces a standard integral. However, it is important to note that this method cannot be applied to all products and on some occasions it may be necessary to apply the method two or more times. The result is given below:

$$\int u \frac{dv}{dx} \, dx = uv - \int v \frac{du}{dx} \, dx$$

The use of this method is illustrated in the following example.

Example:

Find $\int xe^{3x} \, dx$

Solution:

Use integration by parts, with:

$\qquad u = x \qquad\qquad\qquad \dfrac{dv}{dx} = e^{3x}$

$\qquad \dfrac{du}{dx} = 1 \qquad \text{and} \qquad v = \dfrac{1}{3} e^{3x}$

These can then be substituted into the formula to give:

$$\int u \frac{dv}{dx} dx = uv - \int v \frac{du}{dx} dx$$

$$\int xe^{3x} dx = \frac{xe^{3x}}{3} - \int \frac{1}{3} e^{3x} dx$$

$$= \frac{xe^{3x}}{3} - \frac{e^{3x}}{9} + c$$

$$= \frac{e^{3x}}{9} (3x - 1) + c$$

---

integration of trigonometric functions: the integrals of the three trigonometric functions are given below:

$$\int \sin x \, dx = -\cos x + c$$

$$\int \cos x \, dx = \sin x + c$$

$$\int \tan x \, dx = -\ln |\cos x| + c$$

The first two of these results follow from the derivatives of $\sin x$ and $\cos x$, but the third is obtained by using the substitution $u = \cos x$ and noting that

$$\tan x = \frac{\sin x}{\cos x}$$

Note that these results require that x is in radians and not degrees.

Example:

Find the following.

(a) $\int 3\cos (2x) \, dx$ (b) $\int_0^{\pi} \sin \frac{x}{2} \, dx$

Solution:

(a) $\int 3\cos (2x) \, dx = \frac{3}{2} \sin (2x) + c$

(b) $\int_0^{\pi} \sin \frac{x}{2} \, dx = \left[-2 \cos \left(\frac{x}{2} \right) \right]_0^{\pi}$

$$= -2 \cos \left(\frac{\pi}{2} \right) - (-2 \cos (0))$$

$$= 0 + 2$$

$$= 2$$

---

integration using trigonometric substitutions: instead of using the method of integration by substitution with a substitution of the form $u = f(x)$, it is sometimes possible to use a substitution of the form $x = f(\theta)$ where $f(\theta)$ is a trigonometric function. The example below illustrates this process.

Example:

Find $\int \frac{1}{\sqrt{(9 - x^2)}} \, dx$.

Solution:

Using the substitution $x = 3 \sin \theta$, and differentiating gives $\dfrac{dx}{d\theta} = 3 \cos \theta$.

The substitution can then be made into the integral using

$$\int f(x)\, dx = \int F(\theta) \frac{dx}{d\theta}\, d\theta$$

$$\int \frac{1}{\sqrt{(9 - x^2)}}\, dx = \int \frac{1}{\sqrt{(9 - 9\sin^2 \theta)}} \times 3 \cos \theta\, d\theta$$

$$= \int \frac{3 \cos \theta}{\sqrt{(9 \cos^2 \theta)}}\, d\theta$$

$$= \int 1\, d\theta$$

$$= \theta + c$$

$$= \sin^{-1}\left(\frac{x}{3}\right) + c$$

integration of x^n: the integral of x^n is given by

$$\int x^n\, dx = \frac{x^{n+1}}{n+1} + c \qquad \text{if } n \neq -1$$

For the case $n = -1$, the integral becomes

$$\int \frac{1}{x}\, dx = \ln|x| + c$$

Example:

Find:

(a) $\displaystyle\int x^3 + 4x - 5\, dx$ $\qquad$ (b) $\displaystyle\int \frac{1}{x^5}\, dx$ $\qquad$ (c) $\displaystyle\int \frac{4}{x}\, dx$

Solution:

(a) $\displaystyle\int x^3 + 4x - 5\, dx = \frac{x^4}{4} + \frac{4x^2}{2} - 5x + c$

$$= \frac{x^4}{4} + 2x^2 - 5x + c$$

(b) $\displaystyle\int \frac{1}{x^5}\, dx = \int x^{-5}\, dx$

$$= \frac{x^{-4}}{-4} + c$$

$$= \frac{-1}{4x^4} + c$$

(c) $\displaystyle\int \frac{4}{x}\,dx = 4\ln|x| + c$

intercept: an intercept of a line or curve is the point where the curve intersects or crosses an axis. The intercepts are often referred to as the x-intercept and y-intercept. For example, the line $y = 2x - 4$ has y-intercept -4 and x-intercept 2, as shown on the graph below.

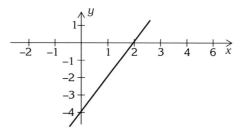

The y-intercept is often referred to as c and is found in the equation of a straight line $y = mx + c$.

If the intercepts of a line are as shown in the diagram below, then its equation can be written as $bx + ay = ab$.

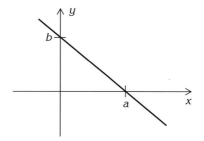

interfering float: when the start time for one activity with a *float* in a *precedence network* affects the float for other activities.

interval bisection: a numerical method used to solve equations of the form $f(x) = 0$. If $f(x)$ is continuous on the interval $a \le x \le b$ and changes sign, then there must be a solution in that interval. Examining the sign of

$$f\left(\frac{a + b}{2}\right)$$

and comparing with the signs of $f(a)$ and $f(b)$, indicates that the solution, x, is in one of the smaller intervals

$$a \le x \le \frac{a + b}{2} \qquad \text{or} \qquad \frac{a + b}{2} \le x \le b$$

This process is repeated with the size of the interval in which the solution lies being halved at each step. The example on page 123 illustrates this process.

Example:

Use interval bisection to find the solution of the equation $x^2 + x - 1 = 0$ that lies in the interval $0 < x < 1$.

Solution:

Let $f(x) = x^2 + x - 1$ and note that $f(0) = -1$ and $f(1) = 1$. As $f(x)$ is continuous and changes sign the equation $f(x) = 0$ must have a solution between 0 and 1.

Now consider the midpoint of this interval, $x = 0.5$. As $f(0.5) = -0.25$, the solution must lie between 0.5 and 1.

Repeating this process leads to the following results (with values of $f(x)$ to 4 d.p.):

$f(0.75) = 0.3125$	so	$0.5 < x < 0.75$
$f(0.625) = 0.0156$	so	$0.5 < x < 0.625$
$f(0.5625) = -0.1211$	so	$0.5625 < x < 0.625$
$f(0.59375) = -0.0537$	so	$0.59375 < x < 0.625$
$f(0.609375) = -0.0193$	so	$0.609375 < x < 0.625$
$f(0.6171875) = -0.0019$	so	$0.6171875 < x < 0.625$
$f(0.62109375) = 0.0069$	so	$0.6171875 < x < 0.62109375$

At this stage it can be concluded that $x = 0.62$ to 2 decimal places.

interval estimation: data from a *random sample* is used to find an interval in which an unknown population *parameter* is expected to lie with a stated degree of confidence. Such intervals are called *confidence intervals*.

interquartile range (IQR): this is a *measure of dispersion*, given by:

$$IQR = Q_3 - Q_1$$

where Q_1 and Q_3 are the upper and lower quartiles.

Example:

Find the interquartile range of this data. A random sample of 20 students was taken and their height in centimeters was recorded:

124	142	181	192	129	141	173	157	125	133
127	196	164	163	164	179	148	143	150	169

Solution:

Place the data into ascending numerical order:

124	125	127	129	133	141	142	143	148	150
157	163	164	164	169	173	179	181	192	196

The lower quartile, Q_1 is the $\dfrac{20 + 1}{4} = 5\frac{1}{4}$th value

$= \dfrac{1}{4}(141 - 133) + 133 = 135$ cm

The upper quartile, Q_3 is the $15\frac{3}{4}$th value

$$= \frac{3}{4}(173 - 169) + 169 = 172 \text{ cm}$$

Therefore the interquartile range, IQR = 172 – 135 = 37 cm

inverse function: if f is a function such that $f(x) = y$, then the inverse of f, denoted f^{-1} is such that $f^{-1}(y) = x$. The domain of the function f becomes the range of the inverse function f^{-1} and the range of f becomes the domain of f^{-1}. The diagram illustrates how f maps x onto y and f^{-1} maps y back onto x.

It is important to note that a function will only have an inverse if the function is a one to one and not a many to one mapping.

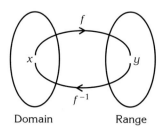

Domain   Range

The graphs of $y = f(x)$ and $y = f^{-1}$ are reflections of each other in the line $y = x$. The graph below illustrates this property, for $f(x) = 2x + 1$ for which

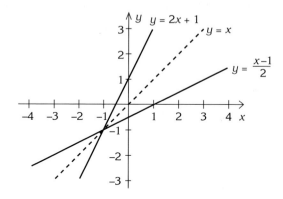

$$f^{-1}(x) = \frac{x - 1}{2}$$

Example:

If $f(x) = \dfrac{1}{1 + x}$ is a function with a domain of real x, such that $0 \le x \le \infty$,

(a) state the range of f
(b) find f^{-1}
(c) state the domain and range of f^{-1}.

Solution:

(a) The range of f is the set of real numbers that satisfy $0 \le y \le 1$.

(b) To find f^{-1}, write $y = \dfrac{1}{1 + x}$ and make x the subject of the expression.

$$y = \frac{1}{1 + x}$$

$$(1 + x)\,y = 1$$

$$y + xy = 1$$

$$xy = 1 - y$$

$$x = \frac{1 - y}{y}$$

So $f^{-1}(x) = \dfrac{1 - x}{x}$

(c) As the range of f is the set of real numbers that satisfy $0 \le y \le 1$ this will be the domain of f^{-1}. The range of f^{-1} will be the set of real numbers that are greater than or equal to zero.

---

irrational numbers: numbers that cannot be written as rational numbers. Some examples of irrational numbers are $\sqrt{2}$, $\sqrt{3}$, π, e, ...

isomorphic graphs: two *graphs* are said to be isomorphic if they have the same number of *vertices* and if the degrees of corresponding pairs of vertices are the same.

isosceles triangle: a triangle with two sides of equal length, with the angles opposite those sides also being equal, as shown below.

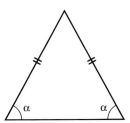

isotype diagram: see *pictogram*.

iterative method: an iterative method produces a sequence of numbers that converge to the solution of an equation. For example, the equation $x^2 = 10$, can be solved using the iterative method:

$$x_{n+1} = \frac{1}{2}\left(x_n + \frac{10}{x_n}\right).$$

Starting with $x_0 = 3$ gives:

$$x_0 = 3, \quad x_1 = 3.16667, \quad x_2 = 3.16228, \quad x_3 = 3.16228$$

The solution of the equation can then be given to 5 decimal places. (See also *general iterative method*.)

joule: the derived SI unit of work and energy, named after the English scientist James Joule (1818–1889).

1 joule (J) = 1 newton meter (N m)

kinetic energy: the kinetic energy of an object is the energy associated with its motion and depends on its mass and its speed. Kinetic energy is a *scalar* quantity.

The kinetic energy of an object modeled as a particle is:

kinetic energy $= \frac{1}{2}mv^2$

where m is the mass of the object and v is its speed.

An object modeled as a rigid body may have kinetic energy due to its angular speed, ω, as well as its linear speed, v. In this case the kinetic energy of a rigid body is given by:

kinetic energy $= \frac{1}{2}mv^2 + \frac{1}{2}I\omega^2$

where I is the moment of inertia of the body about the axis of rotation.

The units of energy are the joule:

1 joule $= 1$ kg m^2 s^{-2} $= 1$ newton meter (N m)

Example:

Find the kinetic energy of:

(a) a 100 meter sprinter of mass 75 kg whose average speed is 9.6 m s^{-1}
(b) a car of mass 1 tonne moving at 96 km h^{-1} (60 mph)

Solution:

(a) $m = 75$, $v = 9.6$, so kinetic energy $= \frac{1}{2} \times 75 \times 9.6^2 = 3456$ J

(b) $m = 1000$, $v = 96$ km h^{-1} $= \dfrac{80}{3}$ m s^{-1} so:

kinetic energy $= \frac{1}{2} \times 1000 \times \left(\dfrac{80}{3}\right)^2 = 355555.\dot{5}$ J

kilogram (kg): the basic SI unit of mass. It is defined as the mass of the prototype cylinder of platinum and iridium, kept in the museum at Sèvres in France.

kinematics: the study of the motion of an object or system of objects without reference to its mass or the forces acting on the object or system. The kinematics of an object or system are thus concerned with position, velocity and acceleration.

Kruskal's algorithm: a *greedy algorithm* for solving *minimum connector problems*.

Step 1: rank the *edges* in order of length.

Step 2: select the shortest edge in the network.

Step 3: select, from the edges that are not in the solution, the shortest edge that does not form a cycle. (Where two edges have the same *weight*, select at random.)

Step 4: repeat step 3 until all the vertices are in the solution.

lamina: a plane body of zero thickness. It is a model used to represent a plane object of uniform thickness which is small compared with the other dimensions of the object. For example, a CD can be modeled as a circular lamina.

Lami's theorem: a little used relationship between the magnitudes and orientations of three concurrent forces in equilibrium. It is a consequence of the sine rule for triangles.

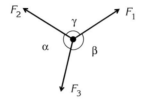

If three forces acting at the point P are in equilibrium then

$$\frac{F_1}{\sin \alpha} = \frac{F_2}{\sin \beta} = \frac{F_3}{\sin \gamma}$$

least squares regression: if n pairs of observations, (x_i, y_i), are plotted on a graph and the points are scattered about a straight line then a *linear relationship* is said to exist between the two variables. The method of least squares fits the "best" line through the points.

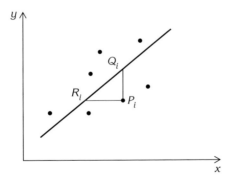

The method achieves this by minimizing the sum of the squared deviations of each point from the "best line."

The line of regression of y on x minimizes $\Sigma(P_i Q_i)^2$ and is used for predicting y values given x values. The line of regression has the equation

$$y - \bar{y} = \frac{S_{xy}}{S_{xx}} (x - \bar{x})$$

where $\bar{y} = \dfrac{\Sigma y_i}{n}$, $\quad \bar{x} = \dfrac{\Sigma x_i}{n}$, $\quad S_{xy} = \Sigma xy - n\bar{x}\bar{y}$, $\quad S_{xx} = \Sigma x^2 - n\bar{x}^2$

The line of regression of x on y minimizes $\Sigma(P_iR_i)^2$ and is used for predicting x values given y values. It has the equation

$$x - \bar{x} = \dfrac{S_{xy}}{S_{yy}}(y - \bar{y}), \quad \text{where} \quad S_{yy} = \Sigma y^2 - n\bar{y}^2$$

Example:

For the following set of data find the least squares regression line of y on x. Use your equation to estimate the value of y for $x = 25$.

x	5	20	45	55	65	80
y	83	79	80	67	58	55

Solution:

The least squares regression line of y on x is given by

$$y - \bar{y} = \dfrac{S_{xy}}{S_{xx}}(x - \bar{x})$$

where $\bar{y} = \dfrac{\Sigma y_i}{n}$, $\quad \bar{x} = \dfrac{\Sigma x_i}{n}$, $\quad S_{xy} = \Sigma xy - n\bar{x}\bar{y}$, $\quad S_{xx} = \Sigma x^2 - n\bar{x}^2$

For the above data $\Sigma x = 270$, $\Sigma y = 422$, $\Sigma xy = 17\,450$, $\Sigma x^2 = 16\,100$, $\Sigma y^2 = 30\,408$ and $n = 6$.

Therefore $\bar{y} = \dfrac{211}{3}$, $\quad \bar{x} = 45$, $\quad S_{xy} = -1540$, and $S_{xx} = 3950$.

$$y - \bar{y} = \dfrac{S_{xy}}{S_{xx}}(x - \bar{x})$$

$$y - \dfrac{211}{3} = \dfrac{-1540}{3950}(x - 45)$$

$$y = -0.39x + 87.88$$

When $x = 25$, $y = -0.39 \times 25 + 87.88 = 78.13$

Most graphic calculators will perform least squares regression on a set of data to produce the regression equation.

light: the term is used to describe objects such as strings, springs, pulleys and rods.

An object is said to be light if its mass is considered to be zero. In practice we would assume that a string, spring, rod, etc. was "light" if its mass is negligible in comparison with other masses in the problem.

limit: The limit of a function $f(x)$ as x approaches a value c is L if the difference between $f(x)$ and L can be reduced by taking a value of x closer to c. Often we are concerned with limits as x approaches 0 or ∞.

For example

$$\lim_{x \to 0} (1 + x) = 1 \quad \text{and} \quad \lim_{x \to \infty} \left(2 + \frac{1}{x} \right) = 2$$

The use of limits is important in the process of *differentiation from first principles*. A sequence can also have a limit. For example the sequence 1, 0.5, 0.25, 0.125, ... has a limit of 0 as the number of terms approaches infinity.

limits of integration: the end points over which a definite integral is to be evaluated. In the expression:

$$\int_a^b f(x) \, dx = [F(x)]_a^b$$

a, b are the limits of the definite integral of $f(x)$.

Example:

Evaluate the integral $\int (x + 1)(2x - 3) \, dx$ between the limits -1 and 2.

Solution:

$$\int_{-1}^{2} (x + 1)(2x - 3) \, dx = \int_{-1}^{2} 2x^2 - x - 3 \, dx$$

$$= \left[\frac{2x^3}{3} - \frac{1}{2}x^2 - 3x \right]_{-1}^{2}$$

$$= \left[\frac{16}{3} - 2 - 6 \right] - \left[-\frac{2}{3} - \frac{1}{2} + 3 \right]$$

$$= -4\frac{1}{2}$$

linear equations: a linear equation is an equation that can be simplified to the form $ax = b$. A simple linear equation is $4x = 20$, but $5(x + 7) = 3(7 - x)$ is also a linear equation that simplifies to $8x = -14$. Linear equations can be solved by carrying out the same operations to both sides of the equations and multiplying out brackets where they appear.

Example:

Solve

(a) $3x - 12 = 21$
(b) $7(x - 6) = 3(x + 9)$

Solution:

(a) To solve $3x - 12 = 21$ first add 12 to both sides of the equation, to give:

$$3x = 33$$

Then divide both sides of the equation by 3, to give the solution:

$$x = \frac{33}{3} = 11$$

(b) To solve the equation $7(x - 6) = 3(x + 9)$, first multiply out the brackets on both sides of the equation, to give:

$$7x - 42 = 3x + 27$$

Then subtract $3x$ from both sides so that x appears only on one side:

$4x - 42 = 27$

Adding 42 to both sides gives:

$4x = 69$

Finally dividing by 4 gives the solution:

$x = \dfrac{69}{4} = 17.25$

---

linear factors: a linear factor is one of the form $ax + b$. The *polynomial* $3x^3 + 20x^2 + 23x - 10$ has the linear factors $(x + 5)$, $(x + 2)$ and $(3x - 1)$, while the polynomial $x^3 + 2x^2 - x - 2$ has one linear factor $(x + 2)$ and one quadratic factor $(x^2 - 1)$.

linear function: a relationship between two variables that can be expressed as an equation and drawn as a straight line.

linear momentum: the *linear momentum* of a particle of mass m moving with velocity $\mathbf{v}$ is a fundamental quantity in mechanics defined by $\mathbf{p} = m\mathbf{v}$. It is a vector quantity in the same direction as the velocity vector.

Linear momentum is an important quantity in mechanics:

• it is conserved in collisions

• the total force acting on a particle equals the rate of change of momentum.

(See also *conservation of momentum*.)

linear programming problem: a problem made up of a linear objective function and constraints formed by linear inequalities.

linear regression: see *least squares regression*.

linear relationship: a linear relationship exists between two variables x and y if $y = ax + b$.

linear transformation: a linear transformation of the x-y plane is the mapping that maps any line in the plane to another line in the plane. The linear transformation T is defined mathematically by

$$\begin{pmatrix} x \\ y \end{pmatrix} \rightarrow T\begin{pmatrix} x \\ y \end{pmatrix} \qquad \text{where } T \text{ is a } 2 \times 2 \text{ nonsingular matrix.}$$

linear velocity: the rate of change of displacement and acts along the tangent to the path or trajectory of a particle's motion. It is what we usually call *velocity*.

local maximum: at a local maximum the gradient of the curve is zero and changes from being positive to negative so that the curve has a peak. The value of the local maximum is not necessarily the maximum value of the curve. The diagram on page 133 shows a curve which has a local maximum, with coordinates (2, 3), whereas the maximum value of the function on the domain (0, 3) is 19.

See *stationary points* for details of how to find a local maximum.

local minimum: at a local minimum the gradient of a curve is zero and changes from being negative to positive, so that the curve has a low point or dip. The value of the local minimum is not necessarily the minimum value of the curve. The curve shown in the graph

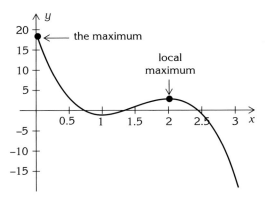

Local maximum

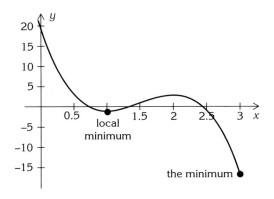

Local minimum

above has a local minimum at the point with coordinates (1, −1) whereas the minimum value of the function on the domain (0, 3) is −17.

See *stationary points* for details of how to find a local minimum.

locus (plural loci): a locus is a set of points that satisfies some specified condition. For example, the diagram here shows the point O and the locus of points that are a distance *d* from O. The locus is a circle of radius *d*.

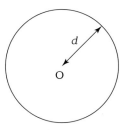

logarithmic differentiation: a process that can be applied to functions that are products, quotients or of the form $y = (f(x))^{g(x)}$. It is most useful for differentiating the last of these. The process is carried out by first taking the natural logarithms of both sides of the expression and simplifying before differentiating.

Example:

Use logarithmic differentiation to find $\dfrac{dy}{dx}$ if:

(a) $y = x^{4x}$
(b) $y = x^{\sin x}$

Solution:

(a) $y = x^{4x}$

First take logs to give:

$$\ln (y) = \ln (x^{4x}) = 4x \ln x$$

Then, differentiating with respect to x using the product rule gives:

$$\frac{1}{y} \times \frac{dy}{dx} = 4x \times \frac{1}{x} + 4 \ln x$$

$$\frac{dy}{dx} = y\left(4x \times \frac{1}{x} + 4 \ln x\right)$$

$$= x^{4x} (4 + 4 \ln x)$$

$$= 4x^{4x} (1 + \ln x)$$

(b) $y = x^{\sin x}$

First take logs of both sides to give:

$$\ln (y) = \ln (x^{\sin x})$$

$$= \sin x \ln x$$

Then differentiate with respect to x using the product rule to give:

$$\frac{1}{y} \times \frac{dy}{dx} = \sin x \times \frac{1}{x} + \cos x \ln x$$

$$\frac{dy}{dx} = y\left(\frac{\sin x}{x} + \cos x \ln x\right)$$

$$= x^{\sin x}\left(\frac{\sin x}{x} + \cos x \ln x\right)$$

logarithmic functions: functions that involve logarithms, for example:

$$f(x) = 4 \log (x + 1).$$

logarithmic plots: when data is collected and displayed using a scatter plot there could be a linear relationship between the two variables and this would be apparent if the plot produced a straight line. However, it may be that a curve is produced. It is possible to reduce some data that gives a curve to a straight line by using logarithms. Two cases that can be treated in this way are if the curve has the equation $y = ax^b$ or $y = ab^x$.

Consider first the relationship $y = ax^b$. If the data satisfies this relationship the scatter plot of the raw data would appear as shown in the graph on the next page, if a is positive.

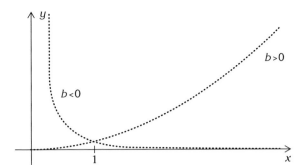

Taking logs of both sides of the relationship gives:

$$\log y = \log (ax^b)$$

$$\log y = \log a + b \log x$$

If $\log y$ is then plotted against $\log x$, a straight line will be produced with vertical intercept of $\log a$ and gradient b.

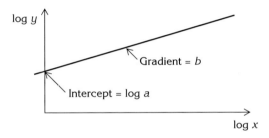

Now consider the relationship $y = ab^x$. If the data satisfies this relationship, the plots of the raw data would look like those shown below, if a is positive.

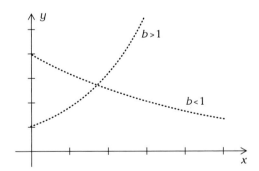

Taking logs of both sides of the relationship gives.

$$\log y = \log (ab^x)$$

$$\log y = \log a + x \log b$$

If $\log y$ is plotted against x, the data will produce a straight line with gradient $\log b$ and vertical intercept $\log a$. This is illustrated on the graph on page 136.

It is possible to obtain logarithmic graph paper for these types of plots.

135

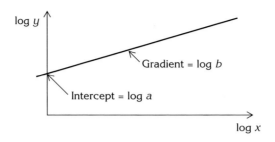

Gradient = log b

Intercept = log a

Example:

Show that the data given in the table satisfies the relationship $y = ax^b$ and find the values of a and b.

x	0	2	4	6	8	10
y	0	1.68	4.80	8.82	13.58	18.97

Solution:

If the relationship $y = ax^b$ holds then a graph of log y against log x will produce a straight line. Taking logs (to base 10) gives the table of values below.

log x	–	0.301	0.602	0.778	0.903	1.000
log y	–	0.225	0.681	0.945	1.133	1.278

Note that it is not possible to find log 0. The graph below shows a plot of log y against log x.

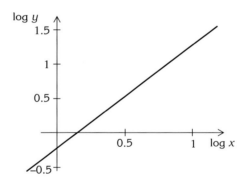

As the plotted points lie on a straight line the relationship holds. The gradient of the line is approximately 1.5 and the vertical intercept –0.23. So $b = 1.5$ and log $a = -0.23$. Then $a = 10^{-0.23} = 0.59$. So the relationship is $y = 0.59x^{1.5}$.

logarithms: if $y = a^x$, then the logarithm to the base a of y is x. This can be written as $\log_a y = \log_a a^x = x$.

The function $y = \log_a x$ is the inverse of the function $y = a^x$. The top graph on page 137 shows the graphs of $y = \log_2 x$ and $y = 2^x$.

Note that $\log_2 x$ is defined only for positive x and that $\log_2 1 = 0$. These properties are true for logarithms to all bases. The second graph on page 137 shows $y = \log_2 x$, $y = \log_4 x$ and $y = \log_{10} x$.

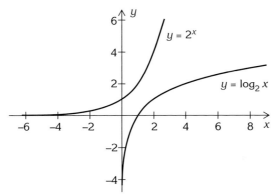

Graphs of $y = \log_2 x$ and $y = 2^x$.

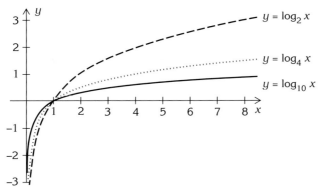

Graphs of $y = \log_2 x$, $y = \log_4 x$ and $y = \log_{10} x$.

Example:

Simplify the following:

(a) $\log_2 32$　　　(b) $\log_9 81$　　　(c) $\log_5 0.2$　　　(d) $\log_7 7$　　　(e) $\log_2 1$

Solution:

(a) $\log_2 32 = \log_2 2^5$
  $= 5$

(b) $\log_9 81 = \log_9 9^2$
  $= 2$

(c) $\log_5 0.2 = \log_5 \tfrac{1}{5}$
  $= \log_5 5^{-1}$
  $= -1$

(d) $\log_7 7 = \log_7 7^1$
  $= 1$

(e) $\log_2 1 = 0$

(See also *rules of logarithms*, *base of a logarithm*, *natural logarithms* and *common logarithms*.)

logarithmic series: there is no simple series for ln x, but a series for ln $(1 + x)$ can be obtained by integrating the binomial expansion for

$$\frac{1}{1 + x}$$

which is valid for $-1 < x < 1$.

The binomial expansion for $\dfrac{1}{1 + x}$ is:

$$\frac{1}{1 + x} = 1 - x + x^2 - x^3 + x^4 - \ldots$$

Integrating both sides of this expression gives:

$$\ln (1 + x) + c = x - \frac{x^2}{2} + \frac{x^3}{3} - \frac{x^4}{4} + \frac{x^5}{5} - \ldots$$

However, as ln $1 = 0$, the constant $c = 0$, so the series is:

$$\ln (1 + x) = x - \frac{x^2}{2} + \frac{x^3}{3} - \frac{x^4}{4} + \frac{x^5}{5} - \ldots$$

The series is valid for $-1 < x < 1$.

Example:

(a) Use the series to find a value for ln(1.1), to 5 decimal places.

(b) Find series for ln $(1 - x)$ and ln $\left(\dfrac{1 + x}{1 - x}\right)$

Solution:

(a) First, note that ln $(1.1) = $ ln $(1 + 0.1)$. Then the series can be applied with $x = 0.1$. This gives:

$$\ln (1 + 0.1) = 0.1 - \frac{0.1^2}{2} + \frac{0.1^3}{3} - \frac{0.1^4}{4} + \frac{0.1^5}{5} - \ldots$$

$$= 0.09531$$

(b) A series for ln $(1 - x)$ can be obtained by substituting $-x$ into the series for ln $(1 + x)$. This gives:

$$\ln (1 - x) = \ln (1 + (-x))$$

$$= (-x) - \frac{(-x)^2}{2} + \frac{(-x)^3}{3} - \frac{(-x)^4}{4} + \frac{(-x)^5}{5} - \ldots$$

$$= -x - \frac{x^2}{2} - \frac{x^3}{3} - \frac{x^4}{4} - \frac{x^5}{5} - \ldots$$

This series is valid for $-1 < x < 1$, as the x term in ln $(1 + x)$ is simply multiplied by -1.

As ln $\left(\dfrac{1 + x}{1 - x}\right) = $ ln $(1 + x) - $ ln $(1 - x)$, a series for ln $\left(\dfrac{1 + x}{1 - x}\right)$ can be obtained by

subtracting the series ln $(1 - x)$ from the series for ln $(1 + x)$. This gives:

$$\ln\left(\frac{1+x}{1-x}\right) = \ln(1+x) - \ln(1-x)$$

$$= \left(x - \frac{x^2}{2} + \frac{x^3}{3} - \frac{x^4}{4} + \frac{x^5}{5} - \ldots\right) - \left(-x - \frac{x^2}{2} - \frac{x^3}{3} - \frac{x^4}{4} - \frac{x^5}{5} - \ldots\right)$$

$$= 2x + \frac{2x^3}{3} + \frac{2x^5}{5} + \ldots .$$

This series is valid for $-1 < x < 1$, as both series are valid for $-1 < x < 1$.

lower bound algorithm: this algorithm is used to find a lower bound to traveling salesperson problems:

Step 1: choose an arbitrary *vertex*. Find the sum of the smallest *arcs* joined to this vertex.

Step 2: ignoring the arbitrary vertex, find a *minimum connector* for this network.

Step 3: add the total from step 1 to the *weight* of the minimum connector in step 2.

A B C D E F G H I J K L M N O P Q R S T U V W X Y Z

mass: the mass of a body is the amount of matter contained in the body.

This may seem a strange definition because we have not defined "matter." The definition of any concept always consists of some more basic concept; for example, linear momentum is defined as "mass times velocity" and we define velocity as rate of change of length. So in these definitions we have used the quantities mass, length and time. Obviously the process cannot go on forever, in that there must be some fundamental concepts. Two of these fundamental concepts are length and time, which one can have a good understanding of, if only from experience. Matter is another fundamental concept, and the definition of mass says that it is a measurable scalar quantity. Note that the fundamental concepts are linked to the three fundamental SI units of meter (length), second (time) and kilogram (mass).

The mass of an object cannot be measured directly, but is obtained using the force of gravity and comparison with "standard" masses. Consider two objects with masses m_1 kg and m_2 kg, respectively. The force of gravity on each object, or weight, is $w_1 = m_1g$ and $w_2 = m_2g$. Hence the ratio of the masses of the two objects is the ratio of their weights.

$$\frac{w_1}{w_2} = \frac{m_1}{m_2}$$

When measured in this way the mass is often called the *gravitational mass*. Because the mass of an object is found by "weighing it" the words weight and mass are often used incorrectly. For example, in everyday speech we say that a bag of sugar weighs one kilogram. We should say that the mass of the bag of sugar is one kilogram or the weight of the bag of sugar is 9.81 newtons. The weight of an object is the force of gravity acting on the object.

It is good practice not to use the word "weight" in mechanics but to use *mass* and *force of gravity* for the gravitational force acting on an object.

An alternative definition of the mass of an object uses *Newton's second law* $F = ma$.

An object of mass m kg requires a force of m newtons to give it an acceleration of 1 m s^{-2}. This is often called *inertial mass*, i.e. the property of the body which is associated with its reluctance to move, or inertia. (The larger the mass of a body, the harder it is to get the body accelerating at 1 m s^{-2}.)

matching: a *graph* that links some of the *vertices* in one subset of a *bipartite graph* to some of the vertices in the other subset.

mathematical induction: see *proof by induction*.

maximum: see *local maximum*.

maximum capacity: the maximum number of "items" that can pass along an *arc* in a *network*. This is represented by a number on the *edge* of the *graph* called its *weight*.

maximum flow/minimum cut algorithm: the maximum *flow* in any *network* equals the value of the *minimum cut* dividing the network.

maximum matching: the optimum solution linking as many pairs of *vertices* as possible in a *bipartite graph*.

mean: the mean is the common term for *arithmetic mean*.

mean deviation (MD): a *measure of dispersion*. It is a measure of the deviation of the data points from the *arithmetic mean* or any other *measure of location*.

The mean deviation from the mean of the n values $x_1, x_2, \ldots x_n$ is:

$$MD = \frac{\Sigma |x - \bar{x}|}{n}$$

The mean deviation from the mean of a frequency distribution is:

$$MD = \frac{\Sigma f |x - \bar{x}|}{\Sigma f}$$

Example:

Find the mean deviation from the mean of the following set of numbers.

5   4   6   8   7   9   5   3   6   8

Solution:

The mean of the numbers is 6.1

x	$x - \bar{x}$	$|x - \bar{x}|$
5	−1.1	1.1
4	−2.1	2.1
6	−0.1	0.1
8	1.9	1.9
7	0.9	0.9
9	2.9	2.9
5	−1.1	1.1
3	−3.1	3.1
6	−0.1	0.1
8	1.9	1.9

$\Sigma |x - \bar{x}| = 15.2$, $n = 10$.   Therefore the mean deviation $= \dfrac{15.2}{10} = 1.52$

measure of dispersion: or *spread* is any statistical measure, such as the *range, interquartile range, standard deviation* or *variance*, which gives a measure of the spread of a data set. Certain measures are more appropriate for certain types of data sets. The best measure of dispersion for *skewed* data is the interquartile range; for *symmetrical* data the best measure is often the variance or standard deviation.

measure of location: any statistical measure, such as the *mean, median* or *mode*, which gives a "typical" or *average* value of the data set. Certain measures are more appropriate for different types of data sets. The best measure of location for *skewed* data is the median; for *symmetrical* data the best measure is the mean.

measure of spread: see *measure of dispersion*.

mechanical energy: kinetic energy and potential energy are two forms of mechanical energy.

(See also *conservation of energy*.)

median: the center or middle item of a data set. To find the median of a set of numbers place the numbers in order of size, locate the middle item and hence identify the median.

Example 1:

Find the median of the following set of numbers:

6   4   6   8   7   9   5   3   7   8   7

Solution:

Placing the numbers in order of size gives

3   4   5   6   6   7   7   7   8   8   9

The middle number is the number which is equidistant from each end, so the median = 7

Example 2:

Find the median of the following set of numbers:

9   1   5   4   0   3   9   4   8   9

Solution:

Placing the numbers in order gives:

0   1   3   4   4   5   8   9   9   9

Here there are an even number of data points and so the median is the midpoint of the middle two values which are 4 and 5. Therefore the median is 4.5.

Example 3:

Estimate the median of the following frequency distribution

Number of take-out meals eaten in last month	Frequency
0	11
1	15
2	17
3	21
4	12
5	9
6	5

Solution:

Number of take-out meals eaten in last month	Frequency	Cumulative frequency
0	11	11
1	15	26
2	17	43
3	21	64
4	12	76
5	9	85
6	5	90

There are 90 observations and so the middle value will be the 45th observation. From the cumulative frequency we can see that the 45th observation will be a 3. Therefore the median number of take-out meals eaten in the last month is 3.

Example 4:

Find the median of the following frequency distribution:

Mass(kg)	Number of students
60–62	4
63–65	9
66–68	20
69–71	13
72–74	4

Solution:

The median can be found either graphically or algebraically.

Graphical method

Draw a *cumulative frequency curve* of the data. Since there are 50 observations, the median must be the 25½th observation. An estimate of the median can be found by reading off the points on the graph.

Upper class boundary	Frequency	Cumulative frequency
59.5	0	0
62.5	4	4
65.5	9	13
68.5	20	33
71.5	13	46
74.5	4	50

It can be estimated from the graph as approximately 67 kg.

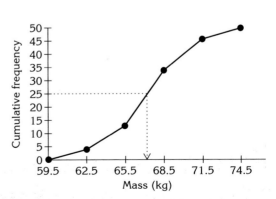

Algebraic method

The median can be estimated algebraically as follows.

The 25½th value lies in the class 65.5–68.5. There are 13 values up to the end of the previous class so 12.5 more are required out of the 20 in the class 65.5–68.5.

The median is given by

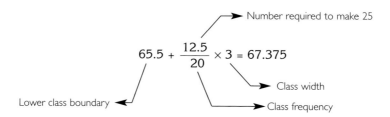

Therefore the median weight is 67.375 kg.

Most graphic calculators will find the median of a set of data.

medians of a triangle: lines that are drawn from one corner to the midpoint of the opposite side. The three medians of a triangle intersect at its *centroid*.

method of least squares: see *least squares regression*

meter: a meter is the fundamental SI unit of length.

metric ton: written tonne, it is a unit of mass equal to 1000 kg.

midpoint: the midpoint of a class is halfway between the two *class boundaries*; i.e.

½ (upper class boundary – lower class boundary)

The midpoint is used when calculating summary statistics from a *frequency table* and, when drawing, *frequency polygons*.

minimum: see *local minimum*.

minimum-connector problem: The *spanning tree* of minimum length for a *network* is a minimum-connector problem.

minimum cut: the cut through a *network* with the minimum capacity.

mode and modal class: The mode of a set of values is the value that occurs most frequently.

Example:

Find the mode of the following data set:

    2   7   2   1   8   2   6   9   10   5   1   4

Solution:

Since 2 occurs most often the mode of the data set is 2.

For a frequency distribution the modal class is the class with the highest *frequency density* (not frequency).

Example:

Find the modal class of the following data

Time (s)	Frequency
$0 \le t < 10$	4
$10 \le t < 20$	7
$20 \le t < 30$	9
$30 \le t < 40$	6
$40 \le t < 50$	5
$50 \le t < 60$	3
$60 \le t < 120$	10

Solution:

We need to find the frequency density $\left(\dfrac{\text{class frequency}}{\text{class width}} \right)$.

Time (s)	Frequency	Frequency density
$0 \le t < 10$	4	0.4
$10 \le t < 20$	7	0.7
$20 \le t < 30$	9	0.9
$30 \le t < 40$	6	0.6
$40 \le t < 50$	5	0.5
$50 \le t < 60$	3	0.3
$60 \le t < 120$	10	0.167

We can see that the modal class is the class $20 \le t < 30$ which has the highest frequency density.

It is possible to estimate the mode from a frequency table. The mode divides the modal class in the same ratio as the increase in frequency density to the decrease in frequency density. It is given by the formula:

$$\text{estimate} = x + \frac{I}{I + D} \times W$$

where x is the lower class bound of the modal class, I is the increase in frequency density, D is the decrease in frequency density and W is the width of the modal class.

Example:

Estimate the mode of the above frequency distribution.

Solution:

The histogram of the data is shown on page 146.

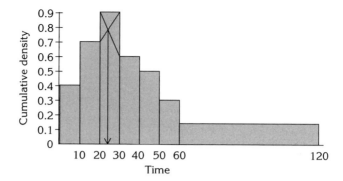

The mode can be estimated as follows. The ratio of the increase in frequency density to decrease in frequency density is 0.2:0.3, (i.e. 2:3). The mode therefore divides the modal class in the ratio of 2:3 and so the mode is 24 seconds.

This can be shown by using the formula:

$$\text{estimate} = x + \frac{I}{I + D} \times W$$

where $x = 20$, $I = 0.2$, $D = 0.3$ and $W = 10$.

Therefore the estimate is

$$20 + \frac{0.2}{0.2 + 0.3} \times 10 = 24 \text{ seconds.}$$

model: a representation of a given situation that can be used to describe the present situation or predict some aspect of the situation in the future. A mathematical model is a representation in the form of a mathematical quantity such as a number, a vector, a formula, an inequality, a graph, a table of values, etc.

For example, *Hooke's law* for an elastic spring $T = ke$ is a model for the tension in the string in terms of its extension. It is an experimental law and assumes that the spring will stretch no matter how small is the force pulling it. For many springs the tension/extension graph will not be the straight line graph through the origin as expected.

A small force is needed before the spring stretches and if the force is too large then a small increase in force can lead to a larger extension than predicted (portion BC of the graph). If the force is too large the spring will break (at point C). Even with a more complicated model, we could use the linear portion AB as a good model in problem solving.

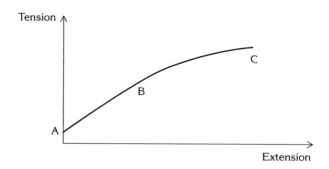

Newton's second law of motion is the fundamental model for Newtonian mechanics and is a relationship between force, mass and acceleration.

Problem solving in mechanics involves using the many models that have been formulated, either as experimental laws (e.g. *Hooke's law*, law of friction, *Newton's law for collisions*) or as mathematical representations (e.g. the use of vectors for forces). It is important to identify the assumptions and simplifications that are involved, before the models can be applied.

modeling: solving a problem described by a realistic situation (such as studying the motion of a bungee jumper) involves some of the following steps:

- understanding the problem
- identifying the important features to be considered
- making some assumptions or simplifications (e.g. a "light, elastic" string)
- defining the variables
- establishing relationships between the variables
- solving the equations
- interpreting and validating the solution
- making improvements to the models
- explaining the outcome.

The process of using these steps to solve a real world problem is called "mathematical modeling." The process has essentially three stages: **formulate** the mathematical model by describing or representing the real world in terms of mathematical symbols and quantities; **solve** any equations that may occur; then use appropriate data to test the goodness of the model and **interpret** the results of the mathematical analysis.

In solving problems in mechanics we use standard models as part of the formulation stage of the modeling process. Each of the standard models is dependent on simplifications and assumptions. The art of good modeling in mechanics is to identify these assumptions/simplifications in each problem.

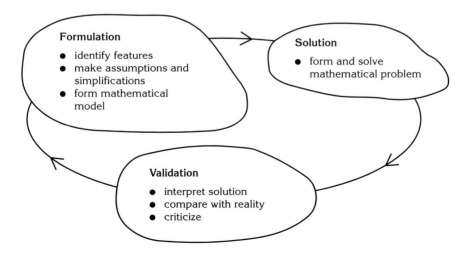

Example:

List the assumptions/simplifications and models in mechanics that would allow us to describe the motion of a bungee jumper.

Solution:

The table shows the assumptions/simplifications that lead to the relevent models in mechanics.

Assumptions/simplifications	Models in mechanics
person is a particle of mass m	Newton's laws, $F = ma$
person drops vertically	one-dimensional motion
perfectly elastic rope	Hooke's law
motion near Earth's surface	force of gravity is constant, mg
no air resistance	only forces are tension and gravity
rope does not break	
initial speed is zero	

modulus: the modulus of a number is an alternative term used to describe the *absolute value* of a number.

modulus of a complex number: the length of the line that would be used to illustrate the *complex number* on an *argand diagram*. The modulus of the complex number $z = a + bi$, is written as $|z|$ and is calculated using Pythagoras' theorem as $|z| = \sqrt{(a^2 + b^2)}$.

moment of a force: The turning effect of a force is called the *moment of the force*. The value of the moment depends on the magnitude of the force and where it is applied.

Consider forces in two dimensions. The magnitude of the moment of a force about a fixed point O is the product of the magnitude of the force and the perpendicular distance d from the point O to the line of action of the force.

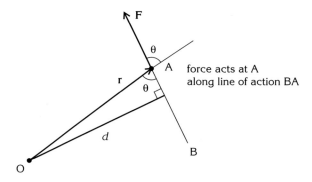

$$\text{moment of } \mathbf{F} \text{ about O} = Fd = Fr \sin \theta$$

where r is the distance OA and θ is the angle between the direction of the force and the direction of OA. ($\mathbf{r}$ is the position vector of A relative to O.)

The moment of a force is a vector quantity with SI units N m. To specify the direction of the moment we use the terms "clockwise" or "counterclockwise." Counterclockwise moments are taken as positive and clockwise moments are taken as negative.

"counterclockwise" moment   "clockwise" moment

The vector sum of the moments of all the external forces acting on an object (the *resultant moment*) equals the rate of change of the angular momentum of the object.

If an object is in equilibrium then the resultant moment about any point is zero.

momentum: see *linear momentum*.

moving averages: given the set of numbers x_1, x_2, ... the moving average of order n is given by the following set of arithmetic means:

$$\frac{x_1 + x_2 + \ldots x_n}{n}, \quad \frac{x_2 + x_3 + \ldots x_{n+1}}{n}, \quad \frac{x_3 + x_4 + \ldots x_{n+2}}{n}, \quad \ldots$$

Moving averages are used in *time series* analysis. The set of moving averages describes the general pattern of the time series.

Example:

Given the numbers 1, 3, 4, 2, 5, 4, 6, 5, find the moving averages of order 4.

Solution:

The first moving average of order 4 is $\dfrac{1 + 3 + 4 + 2}{4} = 2.5$

The second moving average of order 4 is $\dfrac{3 + 4 + 2 + 5}{4} = 3.5$

The third moving average of order 4 is $\dfrac{4 + 2 + 5 + 4}{4} = 3.75$

The fourth moving average of order 4 is $\dfrac{2 + 5 + 4 + 6}{4} = 4.25$

The fifth moving average of order 4 is $\dfrac{5 + 4 + 6 + 5}{4} = 5$

multiplication of complex numbers: the complex numbers $a + bi$ and $c + di$ can be multiplied by expanding the brackets in their product $(a + bi)(c + di)$ and recalling that $i^2 = -1$.

The multiplication of complex numbers can also be treated geometrically, as shown in the diagram here.

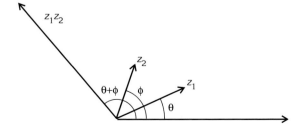

Note that the modulus of the product of two complex numbers is given by $|z_1 z_2| = |z_1| \times |z_2|$ and the argument is given by $\arg(z_1 z_2) = \arg(z_1) + \arg(z_2)$.

Example:

If $z = 4 - 3i$ and $w = -2 - 5i$, find:

(a) z^2

(b) zw

Solution:

(a) $z^2 = (4 - 3i)(4 - 3i)$

$= 16 - 12i - 12i + 9i^2$

$= 16 - 24i + 9 \times (-1)$

$= 16 - 24i - 9$

$= 7 - 24i$

(b) $zw = (4 - 3i)(-2 - 5i)$

$= -8 - 20i + 6i + 15i^2$

$= -8 - 14i + 15 \times (-1)$

$= -8 - 14i - 15$

$= -23 - 14i$

multiplication of polynomials: see *polynomials*.

multiplication of surds: see *surds*.

mutually exclusive events: two events are mutually exclusive if they cannot occur simultaneously. For example, if a die is thrown once, the two different scores, say 1 and 6, cannot occur simultaneously, and therefore they are mutually exclusive events.

If A and B are two mutually exclusive events then $P(A \cap B) = 0$.

Naperian logarithm: is another name for *natural logarithm*.

natural logarithms: logarithms to the base e or approximately 2.7182818. They have the property that $\log_e e^x = x$. It is more common to write ln x rather than $\log_e x$. Natural logarithms can be used to solve equations that contain the *exponential function*.

Example:

Solve the equations:

(a) $10 = 2e^{3x}$
(b) $45 = 15 + 5e^{-0.01x}$

Solution:

(a)
$$
\begin{aligned}
10 &= 2e^{3x} \\
5 &= e^{3x} \\
\ln 5 &= \ln(e^{3x}) \\
\ln 5 &= 3x \\
x &= \frac{\ln 5}{3}
\end{aligned}
$$

(b)
$$
\begin{aligned}
45 &= 15 + 5e^{-0.01x} \\
30 &= 5e^{-0.01x} \\
6 &= e^{-0.01x} \\
\ln 6 &= \ln(e^{-0.01x}) \\
\ln 6 &= -0.01x \\
x &= -\frac{\ln 6}{0.01} \\
&= -100 \ln 6
\end{aligned}
$$

natural numbers: the counting numbers 1, 2, 3, 4, 5, 6, ...

nearest neighbor algorithm: a *greedy algorithm* for finding a *tour* through a *network* by selecting the shortest available *edge* at each stage. The algorithm works by visiting the nearest *vertex* that has not yet been visited until it returns to the starting point.

necessary: a term used to describe a relationship between two statements. When P is a necessary condition for Q, this means that whenever Q is true, then P is also true. We often say Q implies P or write $Q \Rightarrow P$ or $P \Leftarrow Q$. However it does not mean that if P is true then Q is also true, although this may be so in certain cases.

An example considering the convergence of the sum of the terms of a sequence can help to illustrate this idea.

Consider the following two sequences:

$$1, \frac{1}{2}, \frac{1}{3}, \frac{1}{4}, \frac{1}{5}, \dots$$

The terms of this sequence tend to zero, but the sum of the terms is infinite.

$$1, \ \frac{1}{4}, \ \frac{1}{9}, \ \frac{1}{16}, \ \frac{1}{25}, \ \cdots$$

The terms of this sequence also tend to zero, but the sum of the terms is finite and is in fact equal to $\dfrac{\pi^2}{6}$

Let P be the statement that the terms of the sequence tend to zero.

Let Q be the statement that the sum of the terms of the sequence is finite.

It is obvious that P must be true if Q is true, as otherwise the sum of the terms would get bigger and bigger. But if P is true it does not mean that Q must be true.

This is illustrated in the example of the two sequences. For the first one Q is not true but P is true and in the second sequences both P and Q are both true.

Here P is a necessary condition for Q.

See *sufficient* and *necessary and sufficient*.

necessary and sufficient: used when two statements are such that the truth of one implies the truth of the other. For two statements P and Q we can write $P \Leftrightarrow Q$. Sometimes the phrase "P if and only if Q" is used.

Example:

Prove that for positive x and y, $x < y$ if and only if $x^2 < y^2$.

Solution:

First prove that $x < y \Rightarrow x^2 < y^2$

Begin with $x < y$ and multiply by x to give $x^2 < xy$.

Repeat multiplying by y to give $xy < y^2$.

Combining these results gives $x^2 < xy < y^2$ or $x^2 < y^2$, as required.

Next prove that $x^2 < y^2 \Rightarrow x < y$

Begin with $x^2 < y^2$ and subtract x^2 from both sides to give

$0 < y^2 - x^2$ or $0 < (y + x)(y - x)$

But as $x + y > 0$, then $y - x > 0$ or $y > x$, as required.

See also *necessary* and *sufficient*.

negative index: a number raised to a negative power or index is defined as

$$x^{-n} = \frac{1}{x^n}$$

For example: $3^{-2} = \frac{1}{9}$ and $100^{-\frac{1}{2}} = \frac{1}{10}$

network: a *graph* in which every *edge* has a value called its *weight*.

newton (N): the derived unit of force. A force of 1 newton is required to give an object of mass 1 kg an acceleration of $1 \ \text{m s}^{-2}$. The unit is named after the English mathematician and scientist, Sir Isaac Newton (1642–1727). (The force of gravity on a small apple is roughly one newton!)

Newton–Raphson method: this is a powerful numerical method that can be used to find the solutions of a nonlinear equation of the form $f(x) = 0$.

The method states that if x_n is an approximate solution of the equation $f(x) = 0$, then:

$$x_{n+1} = x_n - \frac{f(x_n)}{f'(x_n)}$$

will be a better approximation. This process is repeated until a desired accuracy is reached. If $f'(x_n)$ is close to zero, then the method may break down.

Example:

The equation $x^5 - x - 2 = 0$ has a solution close to 1. Use the Newton–Raphson method to find this solution correct to 4 decimal places.

Solution:

In this example $f(x) = x^5 - x - 2$. Differentiating gives $f'(x) = 5x^4 - 1$.
So the iterative formula will be:

$$x_{n+1} = x_n - \frac{x_n^5 - x_n - 2}{5x_n^4 - 1}$$

Starting with $x_0 = 1$ gives:

$$x_1 = 1 - \frac{1^5 - 1 - 2}{5 \times 1^4 - 1}$$

$$= 1.5$$

$$x_2 = 1.5 - \frac{1.5^5 - 1.5 - 2}{5 \times 1.5^4 - 1}$$

$$= 1.33162$$

Continuing to iterate gives:

$$x_3 = 1.27352$$
$$x_4 = 1.26724$$
$$x_5 = 1.26717$$
$$x_6 = 1.26717$$

So to 4 decimal places the solution is 1.2672.

Newtonian mechanics: a description of the motion of objects based on *Newton's laws of motion*. For these laws to apply, an object is modeled as a particle moving with speeds very much smaller than the speed of light. "Quantum mechanics" deals with the motion of atomic and subatomic particles and "relativistic mechanics" deals with objects moving at speeds close to the speed of light.

Newton's law for collisions: if you drop different types of balls onto the floor so that they bounce then they will rebound to different heights. Newton proposed an experimental law to describe how the impact and rebound speeds are related. He stated that

$$\frac{\text{rebound speed}}{\text{impact speed}} = \text{constant}$$

The ratio is called the *coefficient of restitution*.

Newton's laws of motion: these are the three fundamental laws of motion on which Newtonian mechanics is based and were first published by Isaac Newton in his "Principia Mathematica" of 1687.

First law of motion

In the absence of a resultant force, a particle either stays permanently at rest or moves at constant velocity.

Second law of motion

The resultant external force acting on a particle is proportional to the rate of change of momentum. By choosing the unit of force to be the newton, Newton's second law of motion can be written as

$$\mathbf{F} = \frac{d(mv)}{dt}$$ and if the mass of the particle is constant then $\mathbf{F} = m\mathbf{a}$

where $\mathbf{a}$ is the acceleration of the particle.

Third law of motion

If two particles exert forces on each other, these forces are equal in magnitude and opposite in direction, acting along the line joining the two particles.

Newton's law of gravitation: see *gravitation*.

node (or vertex): a point on a *graph* representing a point where *edges* (arcs) meet.

nonlinear relationships: a relationship between two variables x and y is described as nonlinear if it is not of the form $y = ax + b$. For example, $y = e^x$ and $y = x^2$ are nonlinear relationships.

normal distribution: A discrete random variable X with a *probability density function* of the form

$$f(x) = \frac{1}{\sigma\sqrt{(2\pi)}} \exp\left(-\frac{(x-\mu)^2}{2\sigma^2}\right)$$

is said to have a normal distribution.

μ and σ are the parameters of the distribution and we write $X \sim N(\mu, \sigma^2)$.

The mean of a normal distribution is μ, and the variance is σ^2.

The normal distribution curve is "bell shaped" and is symmetrical about the mean μ. The area under the curve equals 1 (see below).

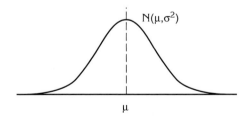

A normal distribution with $\mu = 0$ and $\sigma = 1$ is said to be the "standard normal distribution." The corresponding variable is denoted by Z so that $Z \sim N(0, 1)$.

The shaded area in the diagram below is P(Z < z) = Φ(z).

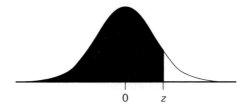

The value of the integral which gives this area is given in "standard normal distribution tables." Most tables are given only for z ≥ 0. Other probabilities can be evaluated by considering the symmetrical properties of the normal curve.

For a ≥ 0, P(Z < a) = Φ(a). This can be found direct from tables, or from a graphic calculator.

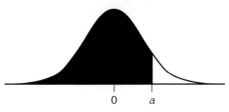

P(Z > a) = 1 − Φ(a), since the total area under the curve equals 1.

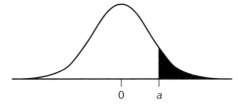

P(Z > −a) = Φ(a)   and   P(Z < −a) = 1 − Φ(a)

For b < c, P(b < Z < c) = Φ(c) − Φ(b)

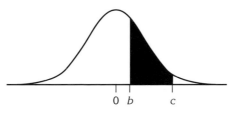

The tables can also be used in reverse to find z if Φ(z) i.e. P(Z < z) is known.

Example 1:

Use standard normal tables to find the values of
(a) $P(Z < 2)$ (c) $P(Z < -1.5)$ (e) $P(-0.5 < Z < 1.5)$
(b) $P(Z > 0.5)$ (d) $P(0.5 < Z < 1.5)$ (f) z if $P(Z < z) = 0.85$

Solution:

Always draw a sketch to help you find the relevant area.

(a)

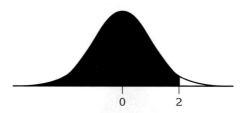

$P(Z < 2) = 0.97725$ direct from tables.

(b)

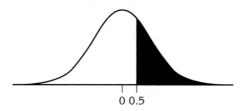

$$P(Z > 0.5) = 1 - P(Z < 0.5)$$
$$= 1 - 0.69146$$
$$= 0.30854$$

(c)

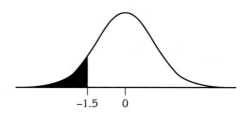

$$P(Z < -1.5) = P(Z > 1.5) = 1 - P(Z < 1.5)$$
$$= 1 - 0.93319$$
$$= 0.06681$$

(d)

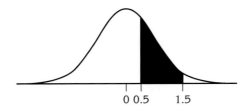

$P(0.5 < Z < 1.5)$ $= P(Z < 1.5) - P(Z < 0.5)$

$= 0.93319 - 0.69146$

$= 0.24173$

(e)

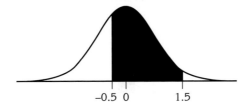

$P(-0.5 < Z < 1.5)$ $= P(Z < 1.5) - P(Z < -0.5)$

$= P(Z < 1.5) - P(Z > 0.5)$

$= P(Z < 1.5) - (1 - P(Z < 0.5))$

$= 0.93319 - (1 - 0.69146)$

$= 0.62465$

(f)

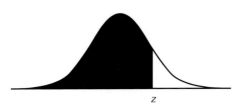

If $P(Z < z) = 0.85$ then z can be found from the tables to be 1.036.

The probabilities associated with any normal distribution, $X \sim N(\mu, \sigma^2)$, can be found by transforming X into the standard normal variable $Z \sim N(0, 1)$.

$$Z = \frac{X - \mu}{\sigma}$$

The probabilities can then be found from the standard normal tables.

Example 2:

Given that $X \sim N(6, 25)$, calculate (a) $P(X > 11)$, (b) $P(5 < X < 7)$.

Solution:

(a) $P(X > 11) = P\left(Z > \dfrac{11 - 6}{5}\right) = P(Z > 1) = 1 - P(Z < 1) = 1 - 0.84134 = 0.15866$

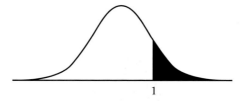

1

(b)

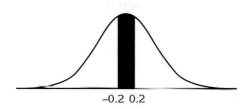

−0.2 0.2

$$P(5 < X < 7) = P\left(\dfrac{5 - 6}{5} < Z < \dfrac{7 - 6}{5}\right) = P(-0.2 < Z < 0.2)$$

$$= P(Z < 0.2) - P(Z < -0.2))$$

$$= P(Z < 0.2) - (1 - P(Z < 0.2))$$

$$= 0.57926 - (1 - 0.57926)$$

$$= 0.15852$$

normal approximation to binomial: the normal distribution can be used to approximate the *binomial distribution* when n is large (> 50) and p is not too small or too large ($0.2 < p < 0.8$). However, if n is very large, the approximation is good whatever the value of p.

If $X \sim \text{Bin}(n, p)$, then X is approximately $N(np, np(1-p))$.

Since the binomial distribution is a *discrete* distribution and the normal distribution is a *continuous* distribution, a continuity correction needs to be applied when using the normal approximation.

The discrete integer (value a) in the binomial distribution becomes the class interval $[(a - 0.5), (a + 0.5)]$ in the normal distribution.

Example:

It is estimated that 70% of students in college take out a student loan. Find the probability that, in a group of 100 students, more than 60 of them have a student loan.

Solution:

Let X be the random variable "number of students with loan." This is a binomial situation with $n = 100$ and p = probability of having a loan = 0.7.

We can approximate $X \sim \text{Bin}(100, 0.7)$ with $X \sim N(100 \times 0.7, 100 \times 0.7 \times 0.3)$ = N(70,21).

We require $P(X > 60.5)$ with the continuity correction applied.

Therefore $P(X > 60.5) \approx P\left(Z > \dfrac{60.5 - 70}{\sqrt{21}}\right) = P(Z > -2.07)$

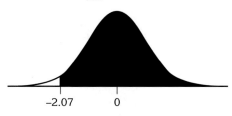

$$-2.07 \qquad 0$$

$$P(Z > -2.07) = P(Z < 2.07) = 0.98077$$

normal approximation to Poisson distribution: the normal distribution can be used to approximate the *Poisson distribution* when μ is large (>20).

If $X \sim \text{Po}(\lambda)$, then X is approximately $N(\lambda, \lambda)$.

Since the Poisson distribution is a *discrete* distribution and the normal distribution is a *continuous* distribution, a continuity correction needs to be applied when using the normal approximation.

The discrete integer value a in the Poisson distribution becomes the class interval $[(a - 0.5),(a + 0.5)]$ in the normal distribution.

Example:

On average a real estate agent sells 25 houses each month. What is the probability that in a given month more than 30 houses are sold?

Solution:

Let X be the number of houses sold. This is a Poisson situation, with $\lambda = 25$. We can approximate $X \sim \text{Po}(25)$ with $N(25, 25)$. We require $P(X > 30.5)$ with the continuity correction applied.

$$P(X > 30.5) \approx P\left(Z > \dfrac{30.5 - 25}{\sqrt{25}}\right) = P(Z > 1.1)$$

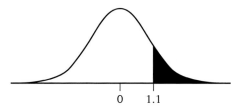

$$0 \qquad 1.1$$

$$P(Z > 1.1) = 1 - P(Z < 1.1) = 1 - 0.8643 = 0.1357.$$

normal reaction: consider two rigid bodies A and B in contact (see the diagram below), then the force exerted by A on B is equal in magnitude and opposite in direction to the force exerted by B on A. Each force can be resolved into two components, one along the tangent to the contact surface and one component along the normal. The tangential force is called the *force of friction* and the normal force is called the *normal reaction*.

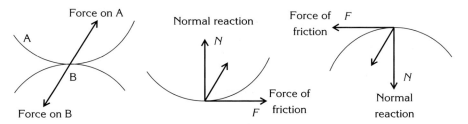

normal to a curve: this is a line that is perpendicular to the tangent to the curve and crosses the curve at the same point that the tangent touches the curve (see below).

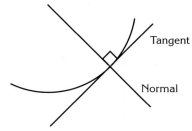

If the gradient of the tangent to the curve is m then the gradient of the normal is $-\dfrac{1}{m}$

Example:

Find the equation of the normal to the curve $y = x^2 + 1$ that passes through the point (3, 10).

Solution:

First differentiate the expression for y to obtain $\dfrac{dy}{dx} = 2x$. So, when $x = 3$, $\dfrac{dy}{dx} = 6$

The gradient of the tangent to the curve at this point is 6 and the gradient of the normal is $-\frac{1}{6}$.

The equation of the normal will be $y = -\dfrac{x}{6} + c$

To find c substitute $x = 3$ and $y = 10$.

$$10 = -\frac{3}{6} + c$$

$$c = 10\tfrac{1}{2}$$

So the equation of the normal is $y = -\dfrac{1}{6}x + 10\tfrac{1}{2}$

null hypothesis H_0: is a statistical hypothesis that can be tested in some way. The null hypothesis is based on the underlying set of assumptions about the population under consideration. The conclusions of a hypothesis test lead to either the acceptance of H_0 or its rejection in favor of the *alternative hypothosis* H_1.

For example, consider a sample of peas from a variety which is known to have a mean mass of 0.1 g. We wish to test if the mean mass of the peas in the sample differs from 0.1 g. Here the null hypothesis is H_0: $\mu = 0.1$ and the alternative hypothesis is H_1 : $\mu \neq 0.1$.

numerator: the numerator of a fraction is the number that appears at the top of the fraction.

For example, the numerator of $\dfrac{a}{b}$ is a. (See also *denominator*.)

A B C D E F G H I J K L M **N** O P Q R S T U V W X Y Z

objective function: the total *cost function* in a linear programming problem that has to be minimized or maximized.

oblique impact: of two bodies occurs when one or both of the bodies has a velocity that is not common to the common normal (or line of centers).

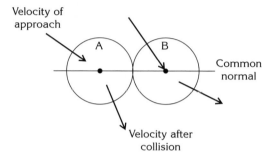

When two smooth bodies collide, the laws governing the collision are:

- conservation of linear momentum
- Newton's experimental law for collisions applied along the line of centers
- the components of the velocities perpendicular to the common normal are unchanged by the impact.

odd function: a function $f(x)$ is an odd function if $f(x) = -f(-x)$. The graph of $y = f(x)$ for an odd function has rotational symmetry about the point (0, 0). Examples of odd functions are $f(x) = x^3$ and $f(x) = \sin x$. The graphs of $y = f(x)$ for these functions are shown below. Note the rotational symmetry.

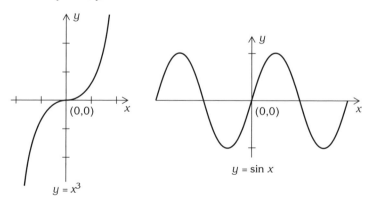

odd vertex: a *vertex* whose degree is odd.

ogive curves: another name for a *cumulative frequency curve.*

one-tailed test: see *alternative hypothosis.*

optimal solution: the best solution according to given conditions.

origin: in *Cartesian coordinates* this is the point where the axes intersect. In general, the origin is the point at which all the coordinates are zero.

oscillations: see *simple harmonic motion* and *damping.*

outcome set: in *probability*, an experiment has a finite number of outcomes called the outcome set *S*.

For example, a die is thrown once. The outcome set S is the set of all possible outcomes, {1,2,3,4,5,6}.

outliers: an outlier is an unusually small or large observation in a set of data. For example, in the following data set the figure 112 is much larger than the rest and so could be classed as an outlier:

37   42   38   56   25   39   40   53   112   47   28

Some calculations based on the data set are more readily affected by outliers than others. For example, as a *measure of location* the *mean* is greatly influenced by an outlier, whereas the *median* is not.

A data point is generally regarded as an outlier if it is more than 1.5× interquartile range beyond the quartiles.

Example:

Identify any outliners in the set of data listed above.

Solution:

For this data set, the lower quartile is 37 and the upper quartile is 53, giving the interquartile range (IQR) as 16.

$$Q_1 - (1.5 \times IQR) = 37 - (1.5 \times 16)$$
$$= 13$$
$$Q_3 + (1.5 \times IQR) = 53 + (1.5 \times 16)$$
$$= 77$$

Any data point less than 13 or greater than 77 will be classified as an outlier. Hence 112 is the only outlier in this data set.

p value: the *p* value obtained when a *hypothesis test* is performed determines the outcome of the test. (The *p* value is usually given when a hypothesis test is performed on a graphic calculator.) If the *p* value is greater that the significance level then the *null hypothesis* H_0 cannot be rejected. If the *p* value is less than the significance level then the null hypothesis H_0 is rejected.

parabola: a *conic section* formed by the intersection of a plane with a cone.

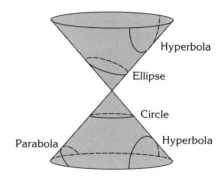

The parametric equations of a parabola are:

$$x = at^2$$
$$y = 2at$$

The Cartesian equation of a parabola is:

$$y^2 = 4ax$$

parallelepiped: a parallelepiped is a solid with six faces, all of which are parallelograms.

parallelogram of forces: a rule used for adding two forces. If two forces $\mathbf{F}_1$ and $\mathbf{F}_2$ are represented in magnitude and direction by the sides AB and AD of a parallelogram then the diagonal AC represents the sum of the two forces. This is called the *resultant force* or *total force* or *net force*.

(See also *vectors*.)

parameter: see *parametric equations*.

parameters of the distribution: all probability distributions have associated parameters which define the distribution. These are summarized for the three main probability distributions, *binomial*, *normal* and *Poisson* distributions as follows:

Distribution	Parameters
Binomial	n, p
Normal	μ, σ^2
Poisson	λ

(See the individual distribution entries for more details.)

parametric differentiation: if expressions for x and y are given in parametric form then it is possible to find $\dfrac{dy}{dx}$ in terms of t. To do this first find $\dfrac{dy}{dt}$ and $\dfrac{dx}{dt}$.

Then $\dfrac{dy}{dx}$ can be found using the result:

$$\frac{dy}{dx} = \frac{\dfrac{dy}{dt}}{\dfrac{dx}{dt}}$$

Example:

If $y = 3t + t^2$ and $x = t + \dfrac{1}{t}$ find $\dfrac{dy}{dx}$

Solution:

First calculate $\dfrac{dy}{dt}$ and $\dfrac{dx}{dt}$

As $y = 3t + t^2$, then $\dfrac{dy}{dt} = 3 + 2t$.

As $x = t + \dfrac{1}{t} = t + t^{-1}$, then $\dfrac{dx}{dt} = 1 - t^{-2} = \dfrac{t^2 - 1}{t^2}$

Finally, $\dfrac{dy}{dx} = \dfrac{\dfrac{dy}{dt}}{\dfrac{dx}{dt}} = \dfrac{(3 + 2t)}{\left(\dfrac{t^2 - 1}{t^2}\right)} = \dfrac{t^2(3 + 2t)}{t^2 - 1}$

parametric equations: a curve $y = f(x)$, can be defined as two parametric equations $y = u(t)$ and $x = v(t)$. The variable t is called a parameter. To plot a curve given the parametric equations, the values of x and y should be calculated for a range of values of t and then plotted. To find the equation of the curve in the form $y = f(x)$, the parameter t must be eliminated from them.

Example:

For the parametric equations

$$y = \frac{2}{t} \text{ and } x = t + 2$$

(a) Find x and y for $t = 1, 2, 3, 4, 5$ and 6, and sketch the curve.
(b) Find y in terms of x.

Solution:

(a) The table below gives the required values.

t	1	2	3	4	5	6
x	3	4	5	6	7	8
y	2	1	$\frac{2}{3} = 0.667$	0.5	0.4	$\frac{1}{3} = 0.333$

The graph below shows these points and a smooth curve drawn through them.

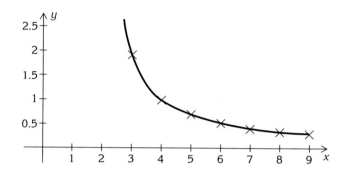

(b) First make t the subject of the equation $x = t + 2$.

$$t = x - 2$$

Then substitute for t in $y = \dfrac{2}{t}$, using this expression, to give.

$$y = \frac{2}{t}$$

$$= \frac{2}{x - 2}$$

---

partial fractions: a way of writing an algebraic fraction in a simpler form, when the denominator can be factorized. There are three main types of partial fraction that can be used if the degree of the denominator is greater than the degree of the numerator.

If the denominator has linear factors then:

$$\frac{ax + b}{(x + p)(x + q)} = \frac{A}{x + p} + \frac{B}{x + q}$$

If the denominator includes a quadratic factor that does not factorize then:

$$\frac{ax^2 + bx + c}{(x + p)(x^2 + q)} = \frac{A}{x + p} + \frac{Bx + C}{x^2 + q}$$

If the denominator contains a repeated linear factor then:

$$\frac{ax^2 + bx + c}{(x + p)(x + q)^2} = \frac{A}{x + p} + \frac{B}{x + q} + \frac{C}{(x + q)^2}$$

If the degree of the denominator is equal to the degree of the numerator a constant term must also be introduced:

$$\frac{ax^2 + bx + c}{(x + p)(x + q)} = A + \frac{B}{x + p} + \frac{C}{x + q}$$

The values of the constants A, B and C can be determined by a number of methods. The examples below illustrate two approaches.

Example 1:

Express $\dfrac{2x^2 - 11}{(x - 2)(x + 1)}$ in partial fractions.

Solution:

The first step is to write the fraction in the form: $A + \dfrac{B}{x - 2} + \dfrac{C}{x + 1}$

The fraction can then be written as:

$$\frac{2x^2 - 11}{(x - 2)(x + 1)} = A + \frac{B}{x - 2} + \frac{C}{x + 1}$$

$$= \frac{A(x - 2)(x + 1) + B(x + 1) + C(x - 2)}{(x - 2)(x + 1)}$$

$$= \frac{Ax^2 + (B + C - A)x + (B - 2A - 2C)}{(x - 2)(x + 1)}$$

Equating coefficients gives:

$A = 2$	$B + C - A = 0$	$B - 2A - 2C = -11$
$B = 2 - C$	$B - 2C = -7$	

Substituting for B in the last equation gives:

$2 - C - 2C = -7$

$3C = 9$ and then $B = 2 - 3$

$C = 3$ $= -1$

So $\dfrac{2x^2 - 11}{(x - 2)(x + 1)} = 2 - \dfrac{1}{x - 2} + \dfrac{3}{x + 1}$

Example 2:

Express $\dfrac{x}{(x - 1)(x + 2)}$ in partial fractions.

Solution:

The partial fractions will be of the form: $\dfrac{A}{x - 1} + \dfrac{B}{x + 2}$

The fraction can then be written as:

$$\frac{x}{(x - 1)(x + 2)} = \frac{A}{x - 1} + \frac{B}{x + 2} = \frac{A(x + 2) + B(x - 1)}{(x - 1)(x + 2)}$$

Now consider the numerators, $x = A(x + 2) + B(x - 1)$.

First, substituting $x = -2$ gives $-2 = -3B$, so $B = \frac{2}{3}$.

Then substituting $x = 1$ gives $1 = 3A$, so $A = \frac{1}{3}$.

So $\dfrac{x}{(x-1)(x+2)} = \dfrac{1}{3(x-1)} + \dfrac{2}{3(x+2)}$

---

particle: the particle forms the fundamental assumption or simplification upon which Newtonian mechanics is based. It is an object with negligible size and internal structure, so that we can represent it mathematically as a point.

The motion of particles is described by *Newton's laws of motion*. In modeling the motion of a real object by the motion of a particle we cannot take into account such properties as spin or the change in orientation of the body with time. However, the particle model will provide a description of the motion of the *center of mass* of a real object.

(See also *rigid body*.)

particular solution of a differential equation: when *initial* or *boundary conditions* are used to determine the values of the constant or constants in a *differential equation*, a particular solution is obtained. This solution would be represented graphically as a single curve, unlike the family of curves that represent the general solution of a differential equation.

Pascal's triangle: given below. It contains many interesting sequences, but is probably most useful for finding the coefficients of x in the *binomial expansion* of $(1 + x)^n$, for positive integer values of n.

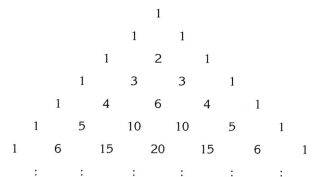

pass: one iteration of an *algorithm*.

path: a route through a *graph* which does not go along any *edge* more than once or visit any *vertex* more than once.

Pearson's product moment correlation coefficient: see *product moment correlation coefficient*.

pendulum: see *simple pendulum*.

percentage point: the *chi squared distribution* is tabulated in percentage points. A percentage point $p\%$, of a χ^2 distribution is that value of χ^2 which has a $p\%$ of the distribution lying to its right. It is written $\chi^2_{p\%}$. See the graph on page 169.

percentage relative numbers: see *index numbers*.

percentiles: $P_1, P_2, \dots, P_{99}$ divide a frequency distribution into 100 equal parts.

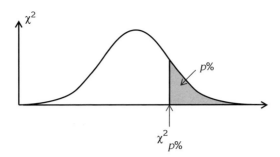

Percentage point

perfect numbers: a perfect number is such that the sum of all its integer divisors, excluding the number itself, is equal to the number. For example, 6 is a perfect number. Its divisors are 1, 2, 3 and 6. Excluding 6 and summing gives $1 + 2 + 3 = 6$. Other perfect numbers include 28 and 496.

perfect squares and cubes: any number that has a square root that is an integer is called a perfect square. For example, 4, 9, 16 and 25 are perfect squares. Similarly, the cube root of a perfect cube is an integer. For example, 8 and 27 are perfect cubes.

perfectly elastic: the term used to describe ideal strings, springs and collisions. A collision is said to be perfectly elastic if the *coefficient of restitution* $e = 1$. For such a collision there is no loss in energy.

A spring or extendible string is said to be perfectly elastic if the tension and extension (or compression for springs) satisfies *Hooke's law*.

period: the period of an oscillation is the time taken for the system to complete one cycle of its motion. For example, for a simple pendulum of length l, the period of small oscillations is $2\pi\sqrt{(l/g)}$.

(See also *simple harmonic motion* and *damping*.)

periodic sequence: a periodic sequence consists of a regularly repeating set of numbers. For example the sequence:

1, 2, 3, 4, 1, 2, 3, 4, 1, 2, 3, 4, ...

is a periodic sequence with a period of 4.

permutations: an ordered arrangement of r objects from n objects. The total number of permutations of r objects from n objects is given by:

$$^{n}P_{r} = \frac{n!}{(n-r)!}$$

Example:

Find the number of permutations of the letters H, I, J, K taken two at a time.

Solution:

$$^{4}P_{2} = \frac{4!}{(4-2)!} = \frac{24}{2} = 12$$

There are 12 permutations of the four letters when taken two at a time.

permutations with one set of identical items: if there are identical items, then the number of permutations of n items taken n at a time when p of the items are identical and the rest are all different is given by

$$\frac{n!}{p!}$$

Example:

In how many ways can the letters of TITLE be arranged?

Solution:

There are five letters so $n = 5$. The T is repeated, twice so $p = 2$.

Therefore the number of permutations is $\dfrac{5!}{2!} = 60$

permutations with two sets of identical items: If there are two sets of identical items then the number of permutations of n items taken n at a time when p of the items are identical and of one kind, and q of the items are identical and of a second kind and the rest are all different is given by:

$$\frac{n!}{p!\,q!}$$

Example:

How many different arrangements of letters can be made using all the letters of the word MINIMUM?

Solution:

There are seven letters so $n = 7$, there are three Ms and so $p = 3$, and two Is so $q = 2$.

Therefore the number of permutations is given by $\dfrac{7!}{3!\,2!} = 420$.

perpendicular bisector: a perpendicular bisector cuts a line into two equal parts and is perpendicular to the line itself. The diagram below shows the perpendicular bisector of AB.

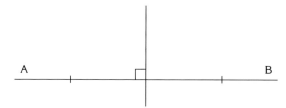

pictogram: a pictogram or *isotype diagram* is the representation of data by symbols, usually pictures. The picture is used to represent a stated number of units of the data, and fractions of this unit are shown as appropriate fractions of the symbol.

Example:

Draw a pictogram of the following data in which the number of each color of cars in a parking lot were recorded.

Color	Number of cars
Red	12
Blue	8
White	7
Green	5
Other	15

Solution:

= 2 cars

Color	Number of cars	
Red	12	
Blue	8	
White	7	
Green	5	
Other	15	

pictorial representation: most data can be represented in some form of picture or diagram. *Bar charts*, *histograms*, *pie charts* and *pictograms* are all ways of displaying data pictorially. A pictorial representation of a data set gives an overall impression of the data. Different types of data sets require different types of pictorial representation.

pie charts: a good way of displaying data that is in a *frequency distribution*. In a pie chart a circle is divided into sectors. The angle of each sector is proportional to the frequency.

The angle is calculated by:

$$\frac{\text{class frequency}}{\text{total frequency}} \times 360°$$

It is useful to display the percentage frequency on the chart.

Example:

Draw a pie chart of the following data in which the number of each color of cars in a parking lot were recorded.

Color	Number of cars
Red	12
Blue	8
White	7
Green	5
Other	15

Solution:

Color of cars	Number	Angle	%
Red	12	$\frac{12}{47} \times 360° = 91.9°$	26%
Blue	8	$\frac{8}{47} \times 360° = 61.3°$	17%
White	7	53.6°	15%
Green	5	38.3°	11%
Other	15	114.9°	32%

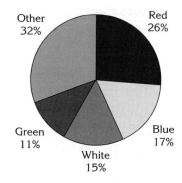

(Note that because of rounding errors the % column does not add to 100%.)

pivot column: the column with the smallest (i.e. "most negative") value in the bottom row of the simplex tableau.

pivot element: the element in the *simplex tableau* that falls in both the *pivot row* and *pivot column*.

pivot row: when the entries in the right-hand column of the *simplex tableau* are divided by the entries in the *pivot column*, the smallest value obtained gives the pivot row.

planar graph: a *graph* that can be drawn in the plane with no *edges* crossing.

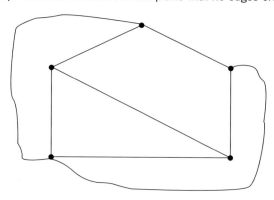

planes: a plane is a flat surface in three-dimensional space. The Cartesian equation of a plane is $ax + by + cz = d$, where the vector

$$\begin{pmatrix} a \\ b \\ c \end{pmatrix}$$

is perpendicular to the plane. Some simple examples are the planes $z = 10$, a horizontal plane that is perpendicular to the vector

$$\begin{pmatrix} 0 \\ 0 \\ 1 \end{pmatrix}$$

at a height of 10 units above the origin and $x + y = 4$, a vertical plane perpendicular to the vector

$$\begin{pmatrix} 1 \\ 1 \\ 0 \end{pmatrix}$$

These planes are illustrated below.

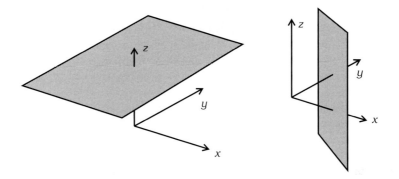

When two planes exist they can either be parallel or intersect to form a line.

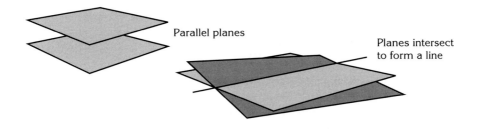

Parallel planes

Planes intersect to form a line

When three planes are considered they can all be parallel, intersect at a point, intersect at a line or pairs of planes can intersect to form three parallel lines.

(See also *vector equation of a plane, angle between two planes* and *angle between a line and a plane*.)

point of inflexion: this is a point on a curve where the tangent to the curve crosses the curve, as shown in the diagrams below.

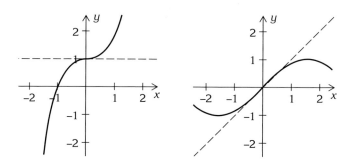

The first graph shows a point of inflexion that is also a "turning" or "stationary point" of the curve. This is the curve $y = x^3 + 1$ which has a point of inflexion at (0,1). The second example, $y = \sin x$, has a point of inflexion at (0,0).

At a point of inflexion the second derivative, d^2y/dx^2, must change sign even though d^2y/dx^2 may not be defined at the point of inflexion. However, just because d^2y/dx^2, is zero this does not necessarily mean that there is a point of inflexion.

Example:

Show that the curve $y = x^3 - 3x^2 + 1$ has a point of inflexion and find its coordinates.

Solution:

Differentiating gives:

$$\frac{dy}{dx} = 3x^2 - 6x \text{ and } \frac{d^2y}{dx^2} = 6x - 6$$

If $\frac{d^2y}{dx^2} = 0$, then $6x - 6 = 0$, so that $x = 1$.

If $x > 1$, then $\frac{d^2y}{dx^2} > 0$. If $x < 1$, then $\frac{d^2y}{dx^2} < 0$

At $x = 1$, $\frac{d^2y}{dx^2}$ changes sign, so the curve has a point of inflection.

Substituting $x = 1$ into the expression for y gives -1, so the coordinates of the point of inflexion are (1,−1).

Poisson approximation to binomial: the Poisson distribution may be used to approximate the *binomial distribution* when n is large (> 50) and p is small (< 0.1).

If $X \sim$ Bin (n, p), then X is approximately Po (np).

Example:

A manufacturing process produces components, of which 0.5% are defective. A sample of 200 components is taken. Find, using a Poisson approximation, the probability that there are two defective components amongst the sample.

Solution:

Let X be the number of defective components. This is a binomial situation with $n = 200$ and p, the probability of being defective $= 0.005$.

We can approximate $X \sim$ Bin $(100, 0.7)$ with $X \sim$ Po $(200 \times 0.005) =$ Po (1).

We require $P(X = 2)$. Therefore $P(X = 2) \approx \dfrac{e^{-1}1^2}{2!} = 0.184$.

Poisson distribution: such a distribution is described by a discrete random variable X with a *probability density function* of the form

$$P(X = x) = \frac{e^{-\lambda}\lambda^x}{x!} \qquad \text{where } x = 0, 1, 2, \ldots$$

λ is the *parameter of the distribution* and we write $X \sim$ Po(λ).

The Poisson distribution is used to estimate probabilities of random events which have a small probability of occurrence. Such events are accident rates, telephone calls arriving at a switchboard, flaws in material, errors in printing, etc.

λ is the mean rate of occurrence per unit time, length, etc.

The mean of a Poisson distribution with parameter λ is λ, and the variance is also λ.

Example:

A student makes an average of three arithmetical errors on each piece of homework. What is the probability that on a particular piece of homework she has made (a) no errors, (b) one error, (c) more than one error?

Solution:

Let the random variable X be "the number of errors per piece of homework." The mean number of errors per piece of homework is the parameter of the distribution and so $\lambda = 3$. Therefore $X \sim$ Po(3).

$$P(X = x) = \frac{e^{-\lambda}\lambda^x}{x!}$$

(a) $P(X = 0) = \dfrac{e^{-3}3^0}{0!} = 0.0498$

(b) $P(X = 1) = \dfrac{e^{-3}3^1}{1!} = 0.149$

(c) $P(X > 1) = 1 - [P(X \le 1)] = 1 - [P(X = 0) + P(X = 1)]$

$$= 1 - (0.0498 + 0.149) = 0.8012$$

Poisson recurrence formula

Poisson recurrence formula: if λ is the parameter of the Poisson distribution, this formula is

$$P(X = x + 1) = \frac{\lambda}{x + 1} P(X = x)$$

This formula allows successive probabilities to be calculated easily once the initial probability is known.

Example:

For a random variable with the distribution $X \sim Po(2)$, use the Poisson recurrence formula to complete the following table.

x	0	1	2	3
$P(X = x)$	0.135			

Solution:

From the Poisson recurrence formula

$$P(X = 1) = \frac{2}{1} P(0) = 2 \times 0.135 = 0.270$$

$$P(X = 2) = \frac{2}{2} P(1) = 1 \times 0.270 = 0.270$$

$$P(X = 3) = \frac{2}{3} P(2) = \frac{2}{3} \times 0.270 = 0.180$$

polar form: a way of writing a complex number in terms of polar *coordinates* (r, θ). It is written in the form:

$$z = r(\cos \theta + i \sin \theta)$$

On the diagram z is represented by the point P or the line OP.

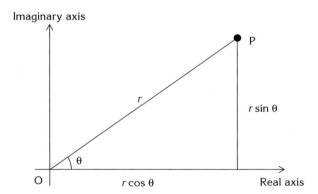

$$|z| = r, \qquad r \geq 0$$
$$\arg z = \theta, \qquad -\pi \leq \theta \leq \pi$$

Example:

Write the complex number $z = 3 + 2i$ in polar form.

Solution:

We need to find the modulus and argument of z.

$$|3 + 2i| = \sqrt{(3^2 + 2^2)} = \sqrt{13}$$

$$\arg(3 + 2i) = \tan^{-1}(\tfrac{2}{3}) = 0.59 \text{ radians}$$

In polar form, $3 + 2i = \sqrt{13}[\cos(0.59) + i\sin(0.59)]$

polygon: a figure with three or more straight sides that do not intersect other than at the vertices. The sum of the interior angles of a polygon with n sides is

$(n - 2) \times 180°$

polygon of forces: when three or more forces acting on an object are represented by a polygon, the sides being defined by the magnitude and directions of the forces, then the line that completes the polygon will be the resultant of all the forces.

Example:

Forces of magnitude 2, 3, 4 and 5 newtons act in the directions shown below; find the resultant force.

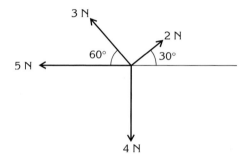

Solution:

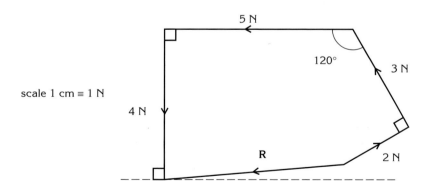

scale 1 cm ≡ 1 N

The figure shows the polygon of forces drawn to scale.

The resultant force **R** has magnitude 4.8 newtons and direction 5° to the horizontal axis in the direction shown in the figure.

In practice, this method is not used to find resultant forces because of the limited accuracy of scale drawing. We would use the algebra of the components of the forces.

(See also *components of vectors*.)

polynomials: a polynomial is an expression of the form:

$$a_n x^n + a_{n-1} x^{n-1} + \ldots + a_1 x + a_0$$

where $a_0, a_1, a_2, \ldots, a_n$ are constants and n is a nonnegative integer. The highest power, n, contained in the polynomial is called its degree. Two examples of polynomials are:

$$x^2 + 6x - 9$$

$$x^4 - 9x^3 + 6x^2 - x + 56$$

When polynomials are being added, the coefficients of like terms must be added. Adding the two polynomials above gives:

$$(x^2 + 6x - 9) + (x^4 - 9x^3 + 6x^2 - x + 56) = x^4 - 9x^3 + (1 + 6)x^2 + (6 - 1)x + (-9 + 56)$$
$$= x^4 - 9x^3 + 7x^2 + 5x + 47$$

When polynomials are being multiplied, every term in each polynomial must be multiplied by every term in the other polynomial. The example below shows a simple case:

$$(x^4 + 4x - 3) \times (x^5 - 2x^3)$$
$$= x^4 \times x^5 + x^4 \times (-2x^3) + 4x \times x^5 + 4x \times (-2x^3) - 3 \times x^5 - 3 \times (-2x^3)$$
$$= x^9 - 2x^7 + 4x^6 - 8x^4 - 3x^5 + 6x^3$$
$$= x^9 - 2x^7 + 4x^6 - 3x^5 - 8x^4 + 6x^3$$

Some polynomials can be factorized; for example:

$$x^4 - 4x^3 - 17x^2 + 60x = x(x - 3)(x + 4)(x - 5)$$

Other polynomials might be factorized, but they will contain nonlinear factors as in the example here:

$$x^4 + 3x^3 + 5x^2 + 6x = x(x + 2)(x^2 + x + 3)$$

The *factor theorem* is very useful when you are trying to factorize polynomials.

If $f(x)$ is a polynomial the number of roots of the equation $f(x) = 0$ is limited by the degree of the polynomial. If the degree is n and n is odd the polynomial can have between 1 and n real roots. If n is even the polynomial can have between 0 and n real roots. The graphs on page 179 show some examples for polynomials of degree 4 and 5.

population: the complete set of objects being studied. A *sample* is any subset of the population.

For example, we might wish to study the height of 18-year-old men in the United States. It would be impractical to measure every member of this population (all 18-year-old men in the United States) and so a sample would be taken.

position vector: the displacement of a point P relative to a fixed origin O is called the *position vector* of P. If the point P has Cartesian coordinates x, y, z then the position vector of P is written as:

$$\mathbf{r} = x\mathbf{i} + y\mathbf{j} + z\mathbf{k}$$

The letter $\mathbf{r}$ is usually reserved for the position vector.

Degree 4

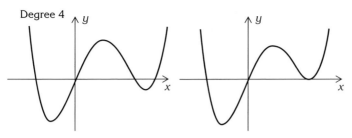

Four real roots

Three real roots one of
which is a repeated root

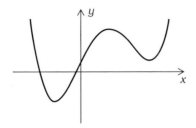

Two real roots

Degree 5

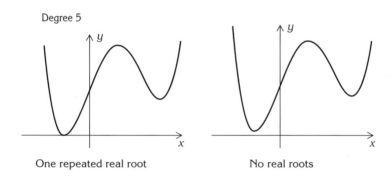

One repeated real root

No real roots

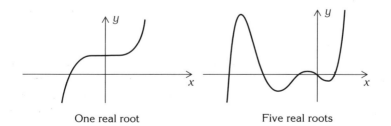

One real root

Five real roots

Polynomials

A
B
C
D
E
F
G
H
I
J
K
L
M
N
O
P
Q
R
S
T
U
V
W
X
Y
Z

potential energy: the mechanical energy that an object has due to its position. The potential energy is the work done by a conservative force in moving from its present position to the chosen position of zero potential energy.

For example, consider the work done by the force of gravity in moving an object from a height h above the origin.

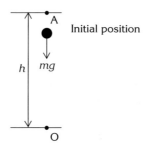

The work done by the force of gravity when the object moves from A to O is

force $\times$ distance $= mgh$

If we choose the zero level of potential energy at the horizontal level through O then we say that the object has a potential energy of mgh at A.

Note that if A was below O then the potential energy would be negative.

Potential energy is a scalar quantity and its unit is the joule (J).

power of a value: see *indices*.

power: defined as the rate at which work is being done. Power is a scalar quantity and its derived SI unit is the watt (W). 1 watt = 1 joule per second. Traditionally the word "horsepower" has been used as a unit of power. One horsepower is equivalent to 746 watts. A Kawasaki Gpz 305 motorcycle is claimed to have a maximum power output of 36 horsepower, which converts to 26 856 watts or almost 27 kW.

For an object moving in a straight line and experiencing a constant force of magnitude F newtons:

$$\text{power} = \text{rate of change of work done} = \frac{\mathrm{d}}{\mathrm{d}t}(Fs) = F\frac{\mathrm{d}s}{\mathrm{d}t} = Fv$$

where v is the speed of the object. This expression of the power, equal to force times speed, is often useful in solving problems.

For example, if the Kawasaki Gpz 305 motorcycle has a maximum speed of 103 mph (46 m s^{-1}) then the forward force when traveling at this speed is 26 856/46 = 584 newtons.

precedence network: a *network* representing the order in which activities need to be carried out to complete a project using *critical path* analysis. When planning a project, the activities required to complete the project are listed, showing which activities must immediately precede others. This is a list of precedence relations. From this list a precedence network is drawn.

precedence relations: the relationships showing the order in which activities must be done.

prime number: a prime number is divisible by only two integers, one and the number itself. The first ten prime numbers are 2, 3, 5, 7, 11, 13, 17, 19, 23 and 29.

Prim's algorithm: an *algorithm* for solving minimum-connector problems.

Step 1: choose a starting *vertex*.

Step 2: join this vertex to the nearest vertex directly connected to it.

Step 3: join the nearest vertex, not already in the solution, to any vertex in the solution, provided that it does not form a cycle.

Step 4: repeat until all vertices have been included.

principal value: when a function produces two or more possible values, a convention is established to define one of these values as the principal value. For example, $\sqrt{4} = 2$ or -2, but the principal value is 2. Principal values are particularly important when working with inverse trigonometric functions, as these are the values given by a calculator.

principle of moments: consider a set of coplanar forces acting on an object.

The vector sum of the forces is the resultant force.

The principle of moments says that the moment of the resultant force about some point in the same plane as the forces, A, is equal to the sum of the moments of the individual forces about A.

This principle is used to find the *center of mass* of an object or composite body.

prism: a prism is a solid with two identical parallel faces and the property that any cross-section parallel to these faces is exactly the same shape as the face. The volume of a prism is equal to the product of the area of cross-section and the length of the prism.

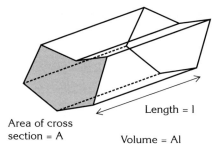

Area of cross
section = A

Length = l

Volume = Al

probability: if an experiment has n equally likely *outcomes* and q of them are the *event* E, then the probability of the event E, P(E), occurring is

$$P(E) = \frac{q}{n}$$

Note: $0 \le P(E) \le 1$.

$P(E) = 0$ means that the event E is an impossibility
$P(E) = 1$ means that the event E is a certainty
$P(\overline{E}) = 1 - P(E)$ means the probability of event E not occurring

Example:

A card is drawn from a normal pack of playing cards. What is the probability that
(a) the card is red?   (b) the card is a club?   (c) the card is a picture?

Solution:

There are 52 cards in a pack and therefore there are 52 equally likely outcomes, $n = 52$.

(a) The event "a red card" leads to 26 possible outcomes; therefore

$$P(\text{red card}) = \frac{26}{52} = \frac{1}{2}$$

(b) The event "a club" leads to 13 possible outcomes; therefore

$$P(\text{club}) = \frac{13}{52} = \frac{1}{4}$$

(c) The event "a picture" leads to 12 possible outcomes; therefore

$$P(\text{picture}) = \frac{12}{52} = \frac{3}{13}$$

The event "a black heart" would have a probability of 0 since it is an impossibility.

The event "either a red card or a black card" would have a probability of 1 since it is a certainty.

probability addition rule: If E_1 and E_2 are two events of the same experiment then the probability of E_1 or E_2 or both occurring is given by:

$$P(E_1 \text{ or } E_2) = P(E_1) + P(E_2) - P(E_1 \text{ and } E_2).$$

In set notation this is written:

$$P(E_1 \cup E_2) = P(E_1) + P(E_2) - P(E_1 \cap E_2).$$

This can be shown on a *Venn diagram*

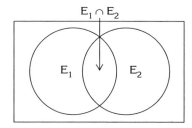

Example:

One card is drawn from a pack of 52 cards. What is the probability that the card is a heart or a six?

Solution:

$$P(\text{heart or six}) = P(\text{heart}) + P(\text{six}) - P(\text{heart and six})$$

$$= \frac{13}{52} + \frac{4}{52} - \frac{1}{52} = \frac{4}{13}$$

probability density function (p.d.f.): the probability density function of a *continuous random variable* or a *discrete random variable*, X, is a function that allocates probabilities to all values that X can take.

For a continuous random variable, X, with a p.d.f. $f(x)$,

$$P(a \leq x \leq b) = \int_a^b f(x)\, dx$$

and $\quad \int_{x_1}^{x_2} f(x)\, dx = 1 \qquad$ where $x_1 \leq x \leq x_2$ is the total range of x.

For a discrete random variable, X, with p.d.f. $P(X = x)$,

$$\sum_{\text{All } x} P(X = x) = 1$$

Example 1:

A continuous random variable, X, has probability density function:

$$f(x) = \begin{cases} kx^2(1-x) & \text{when } 0 \leq x \leq 1 \\ 0 & \text{for all over values of } x \end{cases}$$

(a) Calculate the value of the constant k.
(b) Calculate the probability $P(0.4 \leq x \leq 0.5)$.

Solution:

(a)   We know that $\int_0^1 f(x)dx = 1$

$$\int_0^1 kx^2(1-x)\, dx = k\left[\frac{x^3}{3} - \frac{x^4}{4}\right]_0^1$$

$$= k\left(\frac{1}{3} - \frac{1}{4}\right) = \frac{k}{12}$$

Therefore

$$\frac{k}{12} = 1, \qquad k = 12$$

(b) $P(0.4 \leq x \leq 0.5) = \int_{0.4}^{0.5} 12x^2(1-x)\, dx = 12\left[\frac{x^3}{3} - \frac{x^4}{4}\right]_{0.4}^{0.5}$

$$= 12\left(\frac{0.5^3}{3} - \frac{0.5^4}{4}\right) - 12\left(\frac{0.4^3}{3} - \frac{0.4^4}{4}\right)$$

$$= 0.1333$$

Example 2:

A die whose sides have been renumbered 1, 2, 2, 3, 3, 3, is thrown. Find the probability density function for X, the score obtained.

Solution:

X is a discrete random variable taking on the values 1, 2, 3. Therefore the p.d.f. of X is

x	1	2	3
$P(X = x)$	1/6	1/3	1/2

probability multiplication rule: if E_1 and E_2 are any two events then the probability of both E_1 and E_2 occurring is given by:

$$P(E_1 \text{ and } E_2) = P(E_2) \times P(E_1 | E_2)$$

where $P(E_1 | E_2)$ is the *conditional probability* of E_1 occurring, given that E_2 has occurred.

In set notation this is written:

$$P(E_1 \cap E_2) = P(E_2) \times P(E_1 | E_2)$$

If E_1 and E_2 are *independent* events then:

$$P(E_1 \cap E_2) = P(E_1) \times P(E_2)$$

Example:

One card is drawn from a pack of 52 cards and a fair die is thrown. What is the probability that the card is a heart and the die shows a six.

Solution:

These are independent events therefore

$$P(\text{heart and six thrown}) = P(\text{heart}) \times P(\text{six thrown})$$

$$= \frac{13}{52} \times \frac{1}{6} = \frac{13}{312} = \frac{1}{24}$$

product of factors: when a polynomial is factorized it is written as a product of factors. For example, the polynomial $6x^3 - 57x^2 + 27x$ can be written as a product of factors as $3x(2x - 1)(x - 9)$.

product moment correlation coefficient: a measure of the amount of *correlation* there is between two variables. It is given by:

$$r = \frac{S_{xy}}{\sqrt{(S_{xx} S_{yy})}}$$

where

$$S_{xy} = \Sigma xy - n\bar{x}\bar{y}, \qquad S_{xx} = \Sigma x^2 - n\bar{x}^2, \qquad S_{yy} = \Sigma y^2 - n\bar{y}^2$$

It is sometimes known as "Pearson's product moment correlation coefficient."

When linear regression is performed on a graphic calculator, the correlation coefficient that is calculated is usually Pearson's product moment correlation coefficient.

Example:

Calculate the product moment correlation coefficient for the following data.

x	3	5	7	9	11
y	8	13	15	18	20

Solution:

For this data, $\bar{x} = 7$, $\bar{y} = 14.8$, $\Sigma x^2 = 285$, $\Sigma y^2 = 1182$, $\Sigma xy = 576$ and $n = 5$.

$S_{xy} = 58$, $S_{xx} = 40$, $S_{yy} = 86.8$.

Therefore $r = \dfrac{58}{\sqrt{40 \times 86.8}} = 0.984$ (to 3 d.p.)

product rule: the product rule is used to differentiate expressions that are the product of two functions, such as $x^5 \sin(6x)$ or $e^{5x}\tan(7x)$. The product rule states that, if $y = uv$, then

$$\frac{dy}{dx} = v\frac{du}{dx} + u\frac{dv}{dx}.$$

Example:

Differentiate $y = x^5 \sin(6x)$.

Solution:

Here $u = x^5$ and $v = \sin(6x)$

Differentiating gives

$$\frac{du}{dx} = 5x^4 \quad \text{and} \quad \frac{dv}{dx} = 6\cos(6x).$$

Using the product rule

$$\frac{dy}{dx} = v\frac{du}{dx} + u\frac{dv}{dx}$$

$$\frac{dy}{dx} = \sin(6x) \times 5x^4 + x^5 \times 6\cos(6x)$$

$$= x^4(5\sin(6x) + 6x\cos(6x))$$

projectiles: the problems that look at the motion of objects near to the Earth's surface are called *projectile* problems. In such problems, the *force of gravity* is assumed to be constant with magnitude mg.

For problems in which air resistance is negligible, the path or trajectory of a projectile is a *parabola* with equation in Cartesian coordinates given by:

$$y = h + x\tan\alpha - \frac{gx^2}{2V^2}\sec^2\alpha$$

where V is the launch speed and α is the angle to the horizontal of the initial velocity and the projectile is initially at position $(0, h)$.

proof: a term used to describe a series of logical steps that lead to a conclusion. At each stage of a proof a new statement is derived from a previous one, until the conclusion is reached. The following examples show some simple proofs.

Example:

(a) If $0 < a < 1$, prove that $0 < a^2 < a < 1$.

(b) If $b > 0$, $B > 0$ and $aB < bA$ prove that $\dfrac{a}{b} < \dfrac{a + A}{b + B} < \dfrac{A}{B}$.

Solution:

(a) We are given $0 < a < 1$. As a is positive, multiplying every term in this inequality by a gives the statement below.

$0 \times a < a \times a < 1 \times a$

But $0 \times a = 0$, so

$0 < a^2 < a$.

But $a < 1$, so

$0 < a^2 < a < 1$.

(b) Begin by considering $a(b + B) = ab + aB$ and use $aB < bA$ to obtain

$a(b + B) = ab + aB < ab + bA = b(a + A)$ or $a(b + B) < b(a + A)$

As $b > 0$ and $B > 0$, $b + B > 0$, so we can divide by b and $(b + B)$ without changing

the direction of the inequality, so we obtain $\dfrac{a}{b} < \dfrac{a + A}{b + B}$.

Similarly $A(b + B) = Ab + AB > aB + AB = B(a + A)$ or $A(b + B) > B(a + A)$ and

then dividing by B and $b + B$ gives $\dfrac{a + A}{b + B} < \dfrac{A}{B}$.

Combining these two inequalities gives the required result $\dfrac{a}{b} < \dfrac{a + A}{b + B} < \dfrac{A}{B}$.

See *proof by contradiction, proof by induction, disproof by counter example.*

proof by contradiction: a method of proof. This method can be used when you want to prove that a statement is true. You begin by assuming that the statement is false and show that this is impossible. A classic example of this type of proof is to assume that $\sqrt{2}$ is rational and then to show that this cannot be true. The proof is shown in the example below.

Example:

Prove that $\sqrt{2}$ is irrational.

Solution:

Assume that $\sqrt{2}$ is rational, then it can be expressed in its simplest form as p/q where p and q are natural numbers with no common divisors except 1. Then:

$$\sqrt{2} = \frac{p}{q}$$

$$2 = \frac{p^2}{q^2}$$

$$2q^2 = p^2$$

So p^2 must be even, and p must also be even.

If p is even then we can write $p = 2k$, where k is a natural number, and so $p^2 = 4k^2$.

Combining this with the equation $2q^2 = p^2$ gives

$$2q^2 = p^2$$

$$2q^2 = 4k^2$$

$$q^2 = 2k^2$$

So q^2 is even and then q is also even.

If p and q are both even then the fraction p/q cannot be in its simplest form and so we have a contradiction. This means that the initial assumption that $\sqrt{2}$ is rational is false and so $\sqrt{2}$ must be irrational.

proof by induction: can be applied to prove that a property $P(n)$ is true for all positive integers n. For example it could be used to prove that the sum of the first n positive integers is

$$\frac{n(n + 1)}{2}$$

or that the sum of the squares of the first n positive integers

$$\frac{n(n + 1)(2n + 1)}{6}$$

This proof uses the following argument. First show that the result is true for $n = 1$. Then show that if it is true for any n, then it is also true for $n + 1$, and conclude that the result must hold for all positive integers.

Example:

$$\text{Prove that } \sum_{i = 1}^{n} r^2 = \frac{n(n + 1)(2n + 1)}{6}$$

Solution:

Let $S(n)$ represent the equation $\dfrac{n(n + 1)(2n + 1)}{6} = 1^2 + 2^2 + \ldots + n^2$

First show that $S(1)$ is true.

$$\frac{1 \times (1 + 1)(2 \times 1 + 1)}{6} = \frac{1 \times 2 \times 3}{6} = 1$$

So the result holds for $n = 1$.

Now assume that $S(n)$ is true and use it to verify $S(n + 1)$.

$$1^2 + 2^2 + 3^2 + \ldots n^2 + (n + 1)^2 = \frac{n(n + 1)(2n + 1)}{6} + (n + 1)^2$$

$$= \frac{(n + 1)[n(2n + 1) + 6(n + 1)]}{6}$$

$$= \frac{(n + 1)(2n^2 + 7n + 6)}{6}$$

187

$$= \frac{(n + 1)(n + 2)(2n + 3)}{6}$$

$$= \frac{(n + 1)[(n + 1) + 1][2(n + 1) + 1]}{6}$$

This agrees with the formula given for the sum of $n + 1$ terms rather than the sum of n terms. So $S(n + 1)$ is true if $S(n)$ is true and so by the principle of induction $S(n)$ is true for all n.

proper fractions: an algebraic fraction

$$\frac{p(x)}{q(x)}$$

where $p(x)$ and $q(x)$ are polynomials, is a proper fraction if the degree of $p(x)$ is less than the degree of $q(x)$. For example

$$\frac{x^2 + 8}{x^4 - x^2 + 1}$$

is a proper fraction, but

$$\frac{x^5 + 8}{x^4 - x^2 + 1}$$

is not a proper fraction and is called an *improper fraction*.

Pythagorean identities: the Pythagorean identities are listed below:

$$\sin^2 \theta + \cos^2 \theta = 1$$
$$1 + \tan^2 \theta = \sec^2 \theta,$$
$$1 + \cot^2 \theta = \mathrm{cosec}^2 \theta$$

The first of these identities can be derived by considering a right-angled triangle containing the angle θ, as shown below:

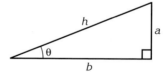

From the triangle

$$\cos \theta = \frac{b}{h} \quad \text{and} \quad \sin \theta = \frac{a}{h}$$

Using these results:

$$\cos^2 \theta + \sin^2 \theta = \frac{b^2}{h^2} + \frac{a^2}{h^2} = \frac{a^2 + b^2}{h^2}$$

Using *Pythagoras' theorem* gives $a^2 + b^2 = h^2$ and so:

$$\cos^2 \theta + \sin^2 \theta = \frac{h^2}{h^2} = 1$$

The other two identities can be derived by dividing each term of this identity by $\cos \theta$ or $\sin \theta$. For example:

$$\frac{\cos^2 \theta}{\cos^2 \theta} + \frac{\sin^2 \theta}{\cos^2 \theta} = \frac{1}{\cos^2 \theta}$$

$$1 + \tan^2 \theta = \sec^2 \theta$$

Pythagoras' theorem: states that, for a right-angled triangle with a *hypotenuse* of length c and the other sides of lengths a and b, then $a^2 + b^2 = c^2$.

quadratic equations: equations with the form $ax^2 + bx + c = 0$. These equations can have two, one repeated or no real solutions. Quadratic equations that have no real solutions do have two complex solutions. Quadratic equations can be solved by three methods: *factorization*, *completing the square* or the quadratic equation formula.

The quadratic equation formula states that, for the equation $ax^2 + bx + c = 0$,

$$x = \frac{-b \pm \sqrt{(b^2 - 4ac)}}{2a}$$

The part of the formula inside the square root is called the discriminant. It is important because it determines the number of real solutions of the equation. There are three important cases:

- $b^2 - 4ac > 0$

  In this case the square root can be found and there are two real solutions.

- $b^2 - 4ac = 0$

  In this case there is only one solution, which is known as a "repeated root."

- $b^2 - 4ac < 0$

  In this case the square root cannot be found and there are no real solutions, but there are two complex solutions.

The graphs below illustrate the three examples.

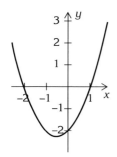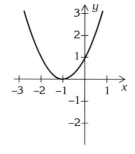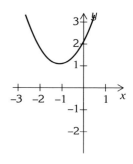

The equations of the curves are:

$$y = x^2 + x - 2 \qquad\qquad y = x^2 + 2x + 1 \qquad\qquad y = x^2 + 2x + 2$$

The values of x for which the curves cross the axes are given by solving the equations below. In this case this is done by factorizing.

$x^2 + x - 2 = 0$	$x^2 + 2x + 1 = 0$	
$(x + 2)(x - 1) = 0$	$(x + 1)(x + 1) = 0$	$x^2 + 2x + 2 = 0$
$x = -2$ or $x = 1$	$x = -1$	
Two real roots	One repeated real root	No real solutions

Example:

(a) Solve the equation $x^2 + 2x - 15 = 0$ by factorizing.

(b) Solve the equation $2x^2 - 5x + 3 = 0$ by using the quadratic formula.

Solution:

(a) Factorizing $x^2 + 2x - 15$ gives:

$$(x - 3)(x + 5) = 0$$

So $x - 3 = 0$ or $x + 5 = 0$

$x = 3$ or $x = -5$

(b) For the equation $2x^2 - 5x + 3 = 0$, $a = 2$, $b = -5$ and $c = 3$. These values can be substituted into the formula to give:

$$x = \frac{-(-5) \pm \sqrt{[(-5)^2 - 4 \times 2 \times 3]}}{2 \times 2}$$

$$= \frac{5 \pm \sqrt{(25 - 24)}}{4}$$

$$= \frac{5 \pm 1}{4}$$

$$= \frac{6}{4} \text{ or } \frac{4}{4}$$

$$= 1.5 \text{ or } 1$$

(See also *complex numbers* and *sum and product of the roots of a quadratic*.)

qualitative data: data which can be assigned qualities or categories are called qualitative data or *categorical data*. It is nonnumerical data. For example, colors, type of accommodation and gender, are all qualitative data types. Qualitative data can be pictorially represented by *bar charts*, *pie charts* and *pictograms*.

quantiles: divide a frequency distribution into equal parts. The *quartiles* divide it into four parts, *deciles* into ten parts and *percentiles* into 100 parts.

quartic: a polynomial of degree 4, such as $x^4 - x^3 + x^2 - 6x + 8$.

quartiles: divide a distribution into four equal parts. The three corresponding values of the variable are denoted by Q_1, the lower quartile, Q_2; the *median*, and Q_3, the upper quartile. The lower quartile has 25% of the distribution below it, the median has 50% below it and the upper quartile, 75% below it (see page 192).

A
B
C
D
E
F
G
H
I
J
K
L
M
N
O
P
Q
R
S
T
U
V
W
X
Y
Z

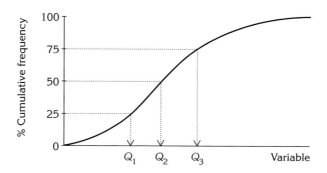

In practice for a sample of size n the lower quartile, Q_1, is the $\frac{(n + 1)}{4}$th value, the median, Q_2, is the $\frac{(n + 1)}{2}$th value

and the upper quartile, Q_3, is the $\frac{3(n + 1)}{4}$th value.

For example, the sample: 3, 5, 9, 10, 11, 12, 16 has 7 values.

The lower quartile is the $\frac{(7 + 1)}{4}$ = 2nd value which is 5.

The median is the $\frac{(7 + 1)}{2}$ = 4th value which is 10.

The upper quartile is the $\frac{3(7 + 1)}{4}$ = 6th value which is 12.

quantitative data: data which is numerical. Such data can either be *discrete* or *continuous* data. For example, consider the data below which comes from a questionnaire.

Question				
Age	25	37	30	...
Gender	m	f	f	...
Height	172	160	170	...
Weight	62	58	72	...
Marital status	s	m	m	...

The responses to the questions about age, height, and weight will give a *quantitative* data set, whereas the other two, gender and marital status, will give a qualitative data set.

Quantitative data can be pictorially represented by *histograms, frequency curves* and *stem and leaf diagrams.*

quicksort algorithm: an *algorithm* for sorting numbers into ascending or descending order. It takes the first number in the list (or the middle number) as a pivot separating the rest of the numbers into two subjects: those that are smaller than the pivot and those that are larger. No reordering of these subsets is carried out. The process is now repeated for each subject and so on.

The first pass using the quicksort algorithm:

5	18	18	18	18	18
18	5	13	13	13	13
13	13	5	26	26	26
26	26	26	5	14	14
3	3	3	3	5	16
14	14	14	14	3	5
16	16	16	16	16	3

quota sampling: in quota sampling, a set number of respondents is required for each category. For example, a researcher who has to interview sixty people may require ten people aged under 18, twenty people aged between 18 and 30, twenty people aged between 30 and 50 and ten people aged over 50.

The advantages of quota sampling are that data is quickly and easily gathered. The disadvantage is that the sample is not random.

quotient rule: this rule is used to differentiate functions like

$$\frac{x^2}{\sin x} \quad \text{or} \quad \frac{\ln x}{x^3}$$

The rule states that:

$$\frac{d}{dx}\left(\frac{u}{v}\right) = \frac{v\dfrac{du}{dx} - u\dfrac{dv}{dx}}{v^2}$$

Example:

Differentiate $\dfrac{x^2}{\sin x}$

Solution:

Here, $u = x^2$ so $\dfrac{du}{dx} = 2x$ and $v = \sin x$ so $\dfrac{dv}{dx} = \cos x$

These can then be substituted into

$$\frac{d}{dx}\left(\frac{u}{v}\right) = \frac{v\dfrac{du}{dx} - u\dfrac{dv}{dx}}{v^2}$$

to give:

$$\frac{d}{dx}\left(\frac{x^2}{\sin x}\right) = \frac{\sin x \times 2x - x^2 \times \cos x}{\sin^2 x}$$

$$= \frac{x(2\sin x - x\cos x)}{\sin^2 x}$$

radians: used to measure the sizes of angles; they are an alternative unit to degrees. One radian is the size of the angle subtended at the center of a circle of radius r by an arc whose length is r.

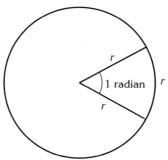

If an angle has size θ radians then the notation is θ rad or θ^c.

To convert radians to degrees (and vice versa) we use the property that a complete turn is 360 degrees or 2π radians. Thus 180 degrees = π radians.

For example, if an angle is given by 35° then in terms of radians:

$$35° = \frac{35}{180} \times \pi = 0.1944 \pi \text{ rad} = 0.61 \text{ rad (to 2 d.p.)}$$

random sampling: in random sampling each member of the population has an equal chance of being selected. The members of the sample are selected by some form of random process, such as through the use of random numbers. Here, each member of the population is allocated a different number. The sample size is determined and a random number generator is used to select the members of the sample.

random variables: see *continuous random variable, discrete random variable*.

range: a *measure of dispersion*. It is the difference between the largest and smallest values.

Range = largest value – smallest value

Example:

Calculate the range of the following numbers:

3, 6, 2, 5, 7, 8, 4, 6, 2, 4, 6, 7, 5, 8, 4.

Solution:

The largest number is 8 and the smallest number is 2. Therefore the range is 8 – 2 = 6.

range of a function: the range of a function is the set that contains the images of all elements of the domain of the function. See *function* for more details.

rate of change: the rate of change of y with respect to x is given by

$$\frac{dy}{dx}$$

Rates of change are usually given with respect to time, t. Information about rates of change is often used to form differential equations. For example, if the rate of change of the size of a population P is proportion to the current population then

$$\frac{dP}{dt} = kP$$

where k is the constant of proportionality.

rational numbers are numbers that can be written in the form

$$\frac{a}{b}$$

where a and b are integers, with $b \neq 0$. Some examples of rational numbers are

$$\frac{1}{7}, 8, -3, -\frac{56}{9}, 1.74...$$

Every terminating or recurring decimal is a rational number and can be written as a fraction.

Example:

Write the following as fractions.

(a) 2.125
(b) 0.234523452345 ...

Solution:

(a) $\quad 2.125 = \dfrac{2125}{1000} = \dfrac{17}{8}$

(b) First, let $x = 0.234523452345$ ...

Then $10\,000x = 2345.234523452345$ ...

Now, subtracting the first of these equations from the second gives:

$10\,000x - x = 2345.234523452345 ... - 0.234523452345$

$9999x = 2345$

$x = \dfrac{2345}{9999}$

rationalization of denominators: the process of changing a fraction such as $\dfrac{a}{\sqrt{b}}$ or $\dfrac{a}{b + c\sqrt{d}}$ so that the denominator does not contain any *surds*.

In the first of these two cases, $\dfrac{a}{\sqrt{b}}$, multiply the top and bottom of the fraction by $\sqrt{b}$.

For the second case, $\dfrac{a}{b + c\sqrt{d}}$, multiply the top and bottom of the fraction by $b - c\sqrt{d}$.

Example:

Rationalize the denominators of:

(a) $\dfrac{2\sqrt{3}}{\sqrt{7}}$ (b)

$\dfrac{5}{4 - 2\sqrt{3}}$

Solution:

(a) $\dfrac{2\sqrt{3}}{\sqrt{7}} = \dfrac{2\sqrt{3} \times \sqrt{7}}{\sqrt{7} \times \sqrt{7}} = \dfrac{2\sqrt{21}}{7}$

(b) $\dfrac{5}{4 - 2\sqrt{3}} = \dfrac{5 \times (4 + 2\sqrt{3})}{(4 - 2\sqrt{3}) \times (4 + 2\sqrt{3})}$

$= \dfrac{20 + 10\sqrt{3}}{16 + 8\sqrt{3} - 8\sqrt{3} - (2\sqrt{3})^2}$

$= \dfrac{20 + 10\sqrt{3}}{16 - 12}$

$= \dfrac{5}{2} (2 + \sqrt{3})$

real numbers: made up of all the *rational* and *irrational numbers*. Some examples are: 0, –8, ¾, 0.783, π, and $\sqrt{7}$.

real part: the real part of a *complex number* is the term that does not contain *i*. For the complex number $z = a + bi$, the real part is *a*. The notation $Re(z) = a$ is often used.

reciprocal: the reciprocal of *x* is $\dfrac{1}{x}$. The reciprocal of the fraction $\dfrac{x}{y}$ is $\dfrac{y}{x}$ as long as x ≠ 0.

rectangular hyperbola: a rectangular *hyperbola* is a hyperbola that has perpendicular *asymptotes*. For example, the curve

$$y = \dfrac{1}{x}$$

is a rectangular hyperbola because it has the axes (which are perpendicular) as asymptotes.

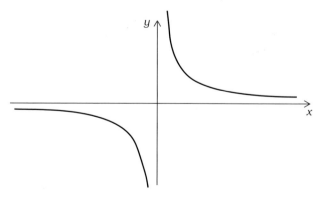

rectilinear motion: this is the motion of an object along a straight line.

recurrence relation: this describes how to obtain the next term of a sequence from its previous term or terms. For example, the recurrence relation

$$u_{n+1} = \frac{2}{3}u_n + 1 \quad \text{with } u_1 = 1$$

generates the sequence

$$1, \frac{5}{3}, \frac{19}{9}, \frac{65}{27}, \ldots$$

regression: the relationship between *random variables*. For a linear relationship between two random variables, the equation of the line of regression can be found by using the method of *least squares regression*.

regular polygon: a polygon in which all the angles are equal and all the sides have the same length.

relative velocity: consider two objects A and B moving with velocities $\mathbf{v}_A$ and $\mathbf{v}_B$, respectively. Then the relative motion of A and B is described in the following way:

velocity of object B relative to A $= \mathbf{v}_{BA} = \mathbf{v}_B - \mathbf{v}_A$

$\mathbf{v}_{BA}$ is called the relative velocity.

Example 1:

A train is traveling along a level straight track at 53 m s^{-1} (roughly 125 mph) in the rain. Drops of rain appear to travel along the window with a speed of 54 m s^{-1} at an angle of 11° to the horizontal. The figure shows the path of a raindrop drawn by a child in the train.

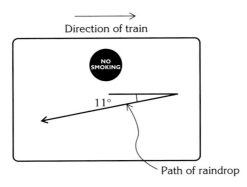

Find the speed and direction of the rain.

Solution:

Choose unit vectors **i** horizontally in the direction of motion of the train and **j** vertically upwards.

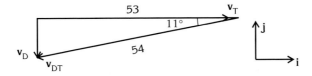

The velocity of the train $\mathbf{v}_T = 53\mathbf{i}$ and the velocity of the raindrop relative to the train is $\mathbf{v}_{DT} = -54\cos(11°)\mathbf{i} - 54\sin(11°)\mathbf{j}$.

If the velocity of a raindrop is $\mathbf{v}_D$, then $\mathbf{v}_{DT} = \mathbf{v}_D - \mathbf{v}_T$, so

$$\mathbf{v}_D = \mathbf{v}_{DT} + \mathbf{v}_T = -54\cos(11°)\mathbf{i} - 54\sin(11°)\mathbf{j} + 53\mathbf{i}$$
$$= -0.0079\mathbf{i} - 10.30\mathbf{j}$$

The speed of the raindrop is $\sqrt{(0.0079^2 + 10.30^2)} = 10.30$ m s^{-1} and the raindrop falls almost vertically.

remainder theorem: the remainder theorem states that if a polynomial $f(x)$ is divided by $(x - a)$, then the remainder is $f(a)$. Note that the *factor theorem* is a special case of the remainder theorem.

Example:

Find the remainder when $x^3 + 4x^2 + 5x + 2$ is divided by $x - 3$.

Solution:

Substituting 3 into the polynomial gives:

$$3^3 + 4 \times 3^2 + 5 \times 3 + 2 = 27 + 36 + 15 + 2$$
$$= 80$$

Note that $x^3 + 4x^2 + 5x + 2 = (x - 3)(x^2 + 7x + 26) + 80$

residuals: When a line of best fit (*least squares regression*) is fitted to a data set, the residuals are the differences between the actual values and the fitted values of the dependent variable. The residuals should always be plotted against the independent variable. If the data does fit a linear relationship then the residual plot will produce a random scatter around the independent axis. Any pattern in the residual plot should lead to the conclusion that the data does not follow a linear relationship.

Example:

For the following data set the least squares regression line of y on x is found to be $y = -0.39x + 87.88$.

x	5	20	45	55	65	80
y	83	79	80	67	58	55

Calculate the residual data set, plot the residuals against x and comment on your plot.

Solution:

For the given x values the fitted y values are calculated using $y = -0.39x + 87.88$.

x	5	20	45	55	65	80
y	83	79	80	67	58	55
fitted y	85.93	80.08	70.33	66.43	62.53	56.68
residuals	-2.93	-1.08	9.67	0.57	-4.53	-1.68

A plot of x against the residuals is as follows:

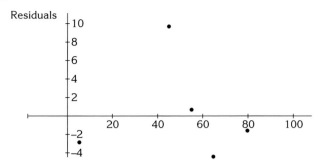

From the plot it can be seen that there is a slight pattern in the residual plot and so maybe a linear relationship is not the best for this data.

resolution of forces: used to find the components of a force in specified directions which are usually perpendicular. (See *components of a force*.)

resource histogram: a diagram to show the number of people needed at any stage of a project being solved by *critical path analysis*.

resource leveling: a system of managing a project to balance the amount of work to be done and the resources available.

restitution: see *coefficient of restitution*.

resultant: the resultant of a vector quantity is the vector sum of two or more quantities. For example, the resultant of two or more forces is the equivalent single force that will provide the same change in motion as the original force system. (See *vector addition*.)

retardation: a word sometimes used to describe the negative acceleration for rectilinear motion when a body is slowing down. This word often causes confusion and its use is best avoided.

rigid body: a model for an object with a shape that cannot be deformed, so that the distance between any two points within the body remains constant, no matter what forces are applied. A rigid body model provides a more general model than the particle model for the motion of real objects, where orientation as well as position of the body are important.

risk paths: *shortest-path problems* involve finding paths with optimal properties not necessarily to do with lengths. If the optimal path is the safest way it is called a risk path.

root: the root of an equation is a term used to describe a value that satisfies the equation. For example, the roots of the equation $x^2 - 3x + 2 = 0$ are $x = 1$ and $x = 2$.

route inspection problems: this involves finding a route through a *network* that passes along every *edge* exactly once and returns to its starting point. This type of problem is often called the "Chinese postman problem" because it was first published by chinese mathematician Mei-Ko Kwan in 1962.

route inspection problem algorithm: an *algorithm* for finding an *optimal solution* to the route inspection is the following:

    Step 1: identify all the odd *vertices* in the *network*.

Step 2: consider all the routes joining pairs of odd vertices and choose the routes of the shortest distance.

Step 3: find the sum of the *weights* on all the *edges* in the network and add the distance found in step 2.

Step 4: find a possible route around the network that repeats the edges identified in step 2 and is of the correct distance.

There may be several possible routes with this length.

rules of logarithms: there are three rules that can be used when manipulating *logarithms*. These are:

$$\log (a \times b) = \log a + \log b$$

$$\log \left(\frac{a}{b} \right) = \log a - \log b$$

$$\log (a^n) = n \log a$$

These rules can be applied to logarithms of any base.

Example:

(a) Write $\log 4 + \log 7 - \log 5$ as a single logarithm.

(b) Write $2 \log x + 3 \log y - \frac{1}{2} \log z$ as a single logarithm.

(c) Expand $\log \left(\dfrac{\sqrt[3]{x}}{y^4} \right)$ in terms of $\log x$ and $\log y$.

Solution:

(a) $\log 4 + \log 7 - \log 5 = \log (4 \times 7) - \log 5$

$$= \log 28 - \log 5$$

$$= \log \left(\frac{28}{5} \right)$$

(b) $2 \log x + 3 \log y - \frac{1}{2} \log z = \log (x^2) + \log (y^3) - \log (z^{1/2})$

$$= \log (x^2 \times y^3) - \log (\sqrt{z})$$

$$= \log \left(\frac{x^2 y^3}{\sqrt{z}} \right)$$

(c) $\log \left(\dfrac{\sqrt[3]{x}}{y^4} \right) = \log(\sqrt[3]{x}) - \log(y^4)$

$$= \log (x^{1/3}) - \log (y^4)$$

$$= \tfrac{1}{3} \log x - 4 \log y$$

sample: a sample is any subset of the population under study. For example, suppose we wished to find the average height of all 18-year-old males in the United States. It would be impractical to measure every 18-year-old male in the US, so a selected number (the sample) are chosen who can be considered representative of the whole population.

There are several ways of selecting samples: see *cluster sampling, quota sampling, random sampling, stratified sampling.*

sampling: the process of obtaining a sample. There are several different sampling methods: *cluster sampling, quota sampling, random sampling, stratified sampling.*

sampling distributions: if a large number of random samples of the same size are taken from the same population then the same statistic (i.e. *mean, variance,* etc.) calculated for all of the samples will form a distribution called the sampling distribution of the statistic.

Example:

A die is thrown four times and the mean score calculated. This is repeated 30 times. The results are shown below:

5.25, 4, 3.75, 4, 3.75, 2.5, 3, 3.5, 2.75, 2, 4, 2.25, 4.25, 4, 3.75, 3.5, 3.75, 4, 4.75, 5, 4.25, 2, 3.75, 2.75, 3.5, 3, 4.25, 3.5, 3.25, 3.75.

Display the sampling distribution of the means in the form of a frequency table.

Solution:

The sampling distribution of the means is as follows:

$\bar{x}$	frequency f
0 – (1)	0
1 – (2)	0
2 – (3)	6
3 – (4)	13
4 – (5)	9
5 – (6)	2

saturated arc: when the flow along an *arc* in a *network* is carrying its *maximum capacity,* it is said to be saturated.

scalar: a scalar is a mathematical object which has the one property of magnitude or size. Physical quantities such as area, volume, mass, temperature, pressure can be modeled by scalars.

scalar product: the product of two vectors that leads to a scalar quantity. (See *dot product.*)

scalene triangle: the lengths of the sides of a scalene triangle are all different. The angles of a scalene triangle are also all different.

scatter diagrams: a scatter diagram is a good way of pictorially representing *bivariate* distributions. One variable is plotted against the other and if there is any relationship between the two variables it can easily be spotted. If the two variables are dependent on each other, the independent variable goes on the horizontal axis and the dependent variable on the vertical axis.

Example:

It is thought that paper tensile strength is dependent on the relative humidity of the atmosphere in which the paper is conditioned. Draw a scatter diagram of the following measurements and comment on any relationship.

Relative humidity (%)	5	20	45	55	65	80
Tensile strength	83	79	73	67	60	55

Solution:

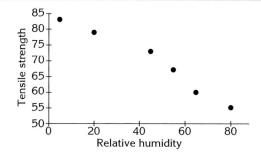

It appears from the scatter diagram that there might be a negative linear relationship between tensile strength and relative humidity.

scientific notation: see *standard form*.

secant: the reciprocal of the *cosine* function and usually written as "sec," So,

$$\sec \theta = \frac{1}{\cos \theta}$$

The graph below shows $y = \cos x$ and $y = \sec x$.

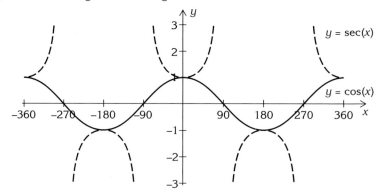

sec⁻¹: this is the inverse of the secant.

second derivative: this is obtained by differentiating the derivative of a function. The second derivative is usually written as:

$$\frac{d^2y}{dx^2} \qquad \text{or} \quad f''(x) \text{ if } y = f(x)$$

The second derivative is often used when determining whether a stationary point is a *local maximum*, *local minimum* or *point of inflexion*.

second derivative test: this is used to determine whether a stationary point is a *local maximum* or a *local minimum*.

If

$$\frac{dy}{dx} = 0 \qquad \text{and} \qquad \frac{d^2y}{dx^2} > 0$$

then the stationary point is a local minimum

If

$$\frac{dy}{dx} = 0 \qquad \text{and} \qquad \frac{d^2y}{dx^2} < 0$$

then the stationary point is a local maximum.

If

$$\frac{dy}{dx} = 0 \qquad \text{and} \qquad \frac{d^2y}{dx^2} = 0$$

then the stationary point could be a local maximum, a local minimum or a point of inflexion. In this case the gradient of the curve either side of the stationary point must be examined.

sector: part of a circle that is enclosed by the circle and two radii. The diagram below shows how a circle is split into two sectors.

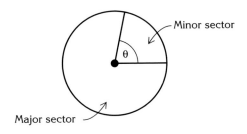

Minor sector

Major sector

The area of the minor sector is:

$$\text{area} = \frac{\theta}{360} \times \pi r^2 \qquad \text{if } \theta \text{ is in degrees}$$

or

$$\text{area} = \frac{1}{2} r^2 \theta \qquad \text{if } \theta \text{ is in radians}$$

segment: a *chord* splits a circle into two segments, as shown below.

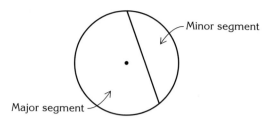

semi-interquartile range: the semi-interquartile range = $\frac{1}{2}(Q_3 - Q_1)$, where Q_1 and Q_3 are the upper and lower *quartiles*.

Example:

Find the semi-interquartile range for the distribution with Q_1 = 137 and Q_3 = 171.

Solution:

Semi-interquartile range = $\frac{1}{2}(171 - 137)$ = 17

sequence: a list of terms that are placed in order according to some rule.

For example the even numbers form the sequence below.

2, 4, 6, 8, 10, …

The square numbers also form a sequence.

1, 4, 9, 16, 25, 36, …

See also the *Fibonacci sequence*.

series: the sum of the terms of a *sequence*. For example:

2 + 4 + 6 + 8 + 10 + …

is a series, while

2, 4, 6, 8, 10, …

is a sequence.

shell sort: an *algorithm* for sorting numbers into ascending or descending order. It involves splitting the numbers into subsets and using shuttle sorts on each subset.

shuttle sort algorithm: an extremely simple and effective sorting method that involves a sequence of swapping neighboring numbers.

First pass: compare the first two numbers and swap if necessary.

Second pass: compare second and third numbers and swap if necessary, then compare first and second numbers and swap if necessary.

Third pass: compare third and fourth numbers and swap if necessary, then compare second and third numbers and swap if necessary, then compare first and second numbers and swap if necessary.

And so on.

shortest path problems: involve finding a path through a network with optimal properties such as the shortest distance, cheapest cost, safest route, etc. *Dijkstra's algorithm* is one *algorithm* for solving such a problem. For example: suppose you live in Mobile and need to travel to Birmingham. What is the shortest route to take?

sigma notation: a way of writing the sum of the terms of a series. The notation

$$\sum_{i=1}^{n} f(i)$$

denotes the sum of the first n terms of the series with $u_i = f(i)$.

For example

$$\sum_{i=1}^{6} 2i = 2 + 4 + 6 + 8 + 10 + 12 = 42$$

and

$$\sum_{i=1}^{5} i^2 = 1 + 4 + 9 + 16 + 25 = 55$$

Two important results are

$$\sum_{i=1}^{n} af(i) = a\sum_{i=1}^{n} f(i) \quad \text{and} \quad \sum_{i=1}^{n} [f(i) + g(i)] = \sum_{i=1}^{n} f(i) + \sum_{i=1}^{n} g(i)$$

significance level: the significance level of a *hypothesis test* gives the probability of wrongly rejecting the true null hypothesis, i.e. the probability of committing a *type I error*.

A significance level of 5% means that, given that the *null hypothesis* is accepted, the probability that an observation will fall in the critical region (thus leading to a rejection of the null hypothesis) is 0.05. The smaller the significance level, the smaller the probability of committing a type I error.

similar: two shapes are said to be similar if corresponding angles are the same in both shapes and the lengths of corresponding pairs of sides are in the same ratio. The diagram below shows two similar pentagons.

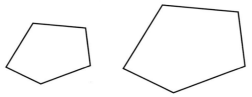

simple harmonic motion: an idealized oscillation in which the amplitude and period remain constant for all time. In practice, oscillations decay due to *damping*.

Mathematically, we define simple harmonic motion as the motion of an object moving in a straight line such that the acceleration is directed towards a fixed point O on the line and has magnitude directly proportional to the distance from the fixed point.

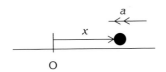

$$\text{acceleration} = \frac{d^2x}{dt^2} = -\omega^2 x$$

where ω is a constant called the angular frequency of the motion and measured in radians per second.

The general solution of this equation that describes the position of the object as a function of time is:

$$x = A\cos(\omega t + \phi)$$

A is called the amplitude of the oscillations, ϕ is the phase and $\frac{2\pi}{\omega}$ is the period.

The frequency of the oscillations is the number of complete cycles per second and given (in Hz) by

$$f = \frac{1}{\text{period}} = \frac{\omega}{2\pi}$$

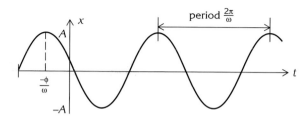

Example:

An object of mass 0.5 kg is attached to the end of a perfect elastic spring of stiffness 15 N m^{-1}. The other end is suspended from the ceiling and the spring/mass system hangs vertically in equilibrium.

The object is pulled down 10 cm from the equilibrium position and released. Describe the subsequent motion of the object.

Solution:

The figure on the next page shows the forces and the displacement from equilibrium x at some general time t. Apply Newton's second law to the object to give

$$0.5g - 15\left(x + \frac{g}{30}\right) = 0.5\frac{d^2x}{dt^2}$$

$$\frac{d^2x}{dt^2} + 30x = 0$$

This is the equation of simple harmonic motion with angular frequency $\sqrt{30}$ rad s^{-1}. Initially, $x = 0.1$ (i.e. 10 cm), giving the solution

$$x = 0.1\cos(\sqrt{30}\,t)$$

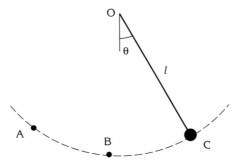

$T = 15(x + \frac{g}{30})$ (Hooke's law)

equilibrium level

x

$0.5g$ (force of gravity)

In the subsequent motion the object oscillates up and down with amplitude 10 cm from the equilibrium level and with period 1.15 seconds.

simple pendulum: a heavy object (called the "bob" and modeled as a particle) suspended from a fixed point by a light rod (assumed to have negligible mass) and allowed to oscillate in a vertical plane. The rod is often replaced by a light inelastic string which must remain taut throughout the motion of the object.

O

θ

l

A   B   C

The object moves back and forth along the arc ABC. The largest value of the angle to the vertical, θ, is the amplitude of the oscillations. If this angle is small then the simple pendulum describes simple harmonic motion with period $2\pi \sqrt{(l/g)}$.

simplex method: the process of using Gaussian elimination steps to solve *linear programming problems*.

simplex tableau: the equations of a *linear programming problem* written in tabular form.

simulation: a process that can be used to test what might happen in situations where an experiment with real subjects may be too long or too dangerous.

simultaneous equations: a pair of equations like
$$2x + 4y = 7$$
$$3x - 12y = -3$$

A solution to these equations will be a value of x and a value of y that satisfy both equations. The equations above are linear simultaneous equations, because they contain only multiples of x and y. Other pairs of equations may not be linear, such as:
$$x^2 + y^2 = 4$$
$$\frac{1}{x} + y = 3$$

There are three methods you are likely to meet that can be used to solve pairs of linear simultaneous equations. These methods are illustrated in the example below.

Example:

Solve
$$2x + 4y = 7$$
$$3x - 12y = -3$$

Solution:

Elimination method

Multiples of each equation are added together so that one variable is eliminated. For example, multiplying the first of the equations above by 3 gives:
$$6x + 12y = 21$$
$$3x - 12y = -3$$

and then adding the two equations gives a linear equation in x that can be solved:
$$9x = 18$$
$$x = 2$$

This value can then be substituted back into one of the original equations to create a linear equation in y, which can then be solved. In this case using the first equation gives:
$$4 + 4y = 7$$
$$4y = 3$$
$$y = \frac{3}{4}$$

Substitution method

First rearrange one of the equations in the form "$y = ...$" and then substitute this expression for y into the other equation. Rearranging the first of the equations above gives:
$$y = \frac{7 - 2x}{4}$$

This is then substituted into the second equation, which becomes:

$$3x - 12\left(\frac{7 - 2x}{4}\right) = -3$$

$$3x - 21 + 6x = -3$$

$$9x = 18$$

$$x = 2$$

This value of x can then be substituted back to find y:

$$y = \frac{7 - 4}{4} = \frac{3}{4}$$

Graphical method

Each of the equations can be considered as the equation of a straight line. The two lines can be drawn and the coordinates of the point of intersection give the solution to the equation. The graph below shows the two lines and the point of intersection.

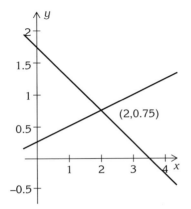

The coordinates give $x = 2$ and $y = 0.75$.

The example considered above has one solution. However it is possible that a pair of linear simultaneous equations has either no solutions or an infinite number of solutions.

The pair of equations:

$$x + y = 2$$

$$x + y = 3$$

has no solution. The graph on page 210 shows that the lines with these equations are parallel and never cross.

The equations:

$$x + y = 3$$

$$3x + 3y = 9$$

are actually the same, as dividing the second equation by 3 would give the first equation. So the solution is represented graphically by a single line. As an infinite number of points lie on that line there are an infinite number of solutions to these equations.

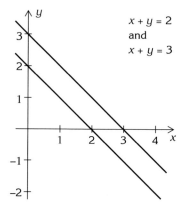

$x + y = 2$
and
$x + y = 3$

Simultaneous equations that are not linear can sometimes be solved by the elimination method, but it is often necessary to use the graphical method. Consider the pair of equations:

$$x + y = 3$$
$$x^2 + y = 9$$

These can be rearranged to give:

$$y = 3 - x$$
$$y = 9 - x^2$$

Drawing these graphs leads to two solutions as shown below.

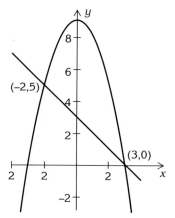

There are two solutions: $x = 3$, $y = 0$ and $x = -2$, $y = 5$.

These equations can also be solved algebraically. Look at the first equation $y = 3 - x$. This can be substituted into the second equation to give:

$$x^2 + 3 - x = 9$$

This quadratic can then be solved by *factorization*:

$$x^2 - x - 6 = 0$$
$$(x - 3)(x + 2) = 0$$

210

$$x - 3 = 0 \quad \text{or} \quad x + 2 = 0$$
$$x = 3 \quad \text{or} \quad x = -2$$

The corresponding y values can then be found by substituting these x values into either of the original equations.

sine:

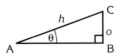

Consider a triangle ABC in which the angle at B is a right angle. If we label the side opposite to angle A as o and the hypotenuse as h, then we define the sine of angle A as

$$\sin \theta = \frac{o}{h}$$

The sine of an angle bigger than 90° is also defined. The graph below shows the sine function $y = \sin(x)$ for angles between 0 and 720°.

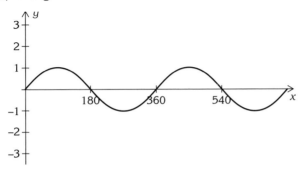

Note that the sine function has a period of 360° (or 2π radians) and that it has an amplitude of 1.

sin⁻¹: this is the inverse of the sine function. It is used when finding an angle given its sine, and is often used in the context of right-angled triangles. When using a calculator you will be given the *principal value*, which will be between −90 and 90 in degrees (or $-\pi/2$ and $\pi/2$ in radians). The graphs below show $y = \sin^{-1} x$ working in degrees on the left and radians on the right. Note that the function is only defined for $-1 \le x \le 1$.

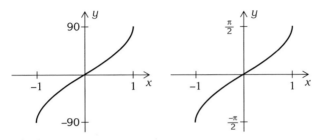

The derivative of $\sin^{-1}$ can be obtained by using the *chain rule* and a *trigonometric identity* in a similar way to the derivative of $\cos^{-1} x$. If $y = \sin^{-1} x$, then

$$\frac{dy}{dx} = \frac{1}{\sqrt{(1 - x^2)}}$$

sine rule: the sine rule can be applied in any triangle to find an unknown angle or length. Before you can apply the sine rule you need to know the lengths of two sides and an opposite angle or the length of one side and two angles.

The sine rule states that:

$$\frac{a}{\sin A} = \frac{b}{\sin B} = \frac{c}{\sin C}$$

It is important to note the convention for labeling the sides and angles as shown in the diagram below.

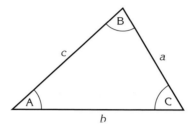

When using the sine rule it is important to be aware that the sine rule can give two possible answers for angles.

Example:

Find the angle marked α in the triangle shown.

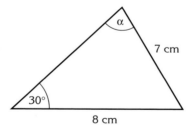

Solution:

Using the sine rule gives

$$\frac{\sin \alpha}{8} = \frac{\sin 30°}{7}$$

$$\sin \alpha = \frac{8 \sin 30°}{7} = \frac{4}{7}$$

$$\alpha = 34.8° \text{ or } 145.2°$$

SI units: The standardized international system of units in general use is called the Système Internationale d'Unités (SI units). The SI units are formed around six base units of which three are relevant to problems in mechanics:

- the meter, m
- the kilogram, kg
- the second, s

These three units form a basis from which all other units in mechanics can be derived using the laws of mechanics.

For example

$$\text{speed} = \frac{\text{distance}}{\text{time}}$$

and so we can define the units of speed as

$$\frac{\text{meter}}{\text{second}} = \text{m s}^{-1}$$

For force we can use Newton's second law, $F = ma$, from which we deduce that:

units of force = units of mass × units of accleration = kg m s^{-2}

There are many such derived units and they often have special names. For example, the unit of force is also called the newton. However, these special names can always be written in terms of the base units: 1 N = 1 kg m s^{-2}

The SI unit of angle is the radian and this is called a supplementary unit because angle is a dimensionless quantity.

sink: the finishing point of a directed *network*.

skewness: refers to the symmetry in the shape of a frequency distribution.

A distribution is symmetrical if the difference between the mean and the median (mean–median) is zero. An appropriate *pictorial representation* of the data (*histogram, stem and leaf diagram*, etc.) would produce a mirror image about the center, as shown below.

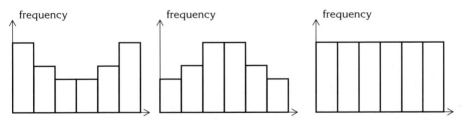

A distribution is positively skewed (or skewed to the right) if the mean minus the median is greater than zero. Such data, when represented by a histogram, would have a "right tail" that is longer than the "left tail," as shown below.

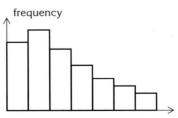

A distribution is negatively skewed (or skewed to the left) if the mean minus the median is less than zero. Such data when represented by a histogram would have a "left tail" that is longer than the "right tail," as shown below.

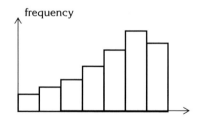

If data is skewed, then the best *measure of location* is the median and the best *measure of dispersion* is the interquartile range.

If data is symmetrical, the best *measure of location* is the mean and the best *measure of dispersion* is the standard deviation.

slack variables: the variables introduced in *linear programming problems* which change the constraints from inequalities into equations. The fundamental theorem of linear programming says that the solution of the problem occurs at one of the corners of the intersection of the planes forming the feasible region.

slant height: the slant height of a cone is the distance from the top of the cone to the base, measured on the surface of the cone.

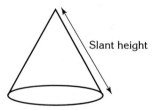

Slant height

slope: a term that is sometimes used instead of *gradient*.

smooth surface: a surface for which it is assumed that the frictional force is negligible.

solving linear inequalities: linear inequalities can be solved in much the same way as a linear equation by carrying out the same operation on both sides of the inequality, but care must be taken when multiplying or dividing by a negative number. When multiplying or dividing by a negative number the direction of the inequality is reversed. Consider the following simple example.

$$9 > 2$$

Multiplying both sides by −2 gives:

$$-18 > -4$$

Clearly this is not true. However reversing the direction of the inequality gives the correct statement:

$$-18 < -4$$

In practice it is best to avoid multiplying or dividing by a negative number.

Example:

Solve the following inequalities:

(a) $3x - 5 \geq -11$, (b) $3 - 2x \leq 12 + 4x$, (c) $7 - 2x > 3$

Solution:

(a) To solve this inequality add 5 to both sides and then divide by 3 as shown below:

$$3x - 5 \geq -11$$
$$3x \geq -6$$
$$x \geq -2$$

(b) The first stage is to add $2x$ to both sides of the equation:

$$3 - 2x \leq 12 + 4x$$
$$3 \leq 12 + 6x$$

Then 12 can be subtracted from both sides and then dividing by 6 gives the solution:

$$-9 \leq 6x$$
$$-\frac{9}{6} \leq x \qquad \text{or} \qquad x \geq -\frac{3}{2}$$

(c) First add $-2x$ to both sides of the inequality so that there is no need to divide by a negative number later in the solution.

$$7 - 2x > 3$$
$$7 > 3 + 2x$$

Then subtracting 3 and dividing by 2 gives:

$$7 > 3 + 2x$$
$$4 > 2x$$
$$2 > x \qquad \text{or} \qquad x < 2$$

solving quadratic inequalities: quadratic inequalities can be solved by two main methods.

Graphical method

The graph below shows the graph of $y = x^2 + 2x - 8$.

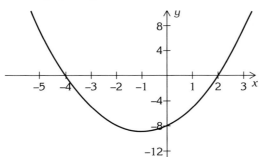

From the graph the solution to the required inequality can be found. For example:

the solution of the inequality $x^2 + 2x - 8 \leq 0$ is $-4 \leq x \leq 2$

the solution of the inequality $x^2 + 2x - 8 > 0$ is $x < -4$ or $x > 2$

Factorization method

Factorize the quadratic and examine the signs of the factors. The following example illustrates this method, for the inequality $x^2 + 7x + 10 \leq 0$.

Factorizing the inequality gives:

$$x^2 + 7x + 10 \leq 0$$

$$(x + 2)(x + 5) \leq 0$$

These factors change sign when $x = -2$ and $x = -5$. These values form the basis for creating the table below.

	$x < -5$	$-5 < x < -2$	$x > -2$
$(x + 2)$	negative	negative	positive
$(x + 5)$	negative	positive	positive
$(x + 2)(x + 5)$	positive	negative	positive

Note that when $(x + 2)$ and $(x + 5)$ are both positive or both negative, then $(x + 2)(x + 5)$ is positive and that when $(x + 2)$ and $(x + 5)$ have different signs, then $(x + 2)(x + 5)$ is negative. So the solution to the inequality $x^2 + 7x + 10 \geq 0$ is $-2 < x < -5$.

solving trigonometric equations: trigonometric equations will have more than one solution, because of the periodic nature of the functions. The graph below illustrates some of the solutions of the equation:

$$\cos x = \frac{1}{2}$$

These solutions are $-420°$, $-300°$, $-60°$, $60°$, $300°$ and $420°$. There are clearly many other solutions. The general solution of this equation is $x = \pm 60 + 360n$, where n is an integer.

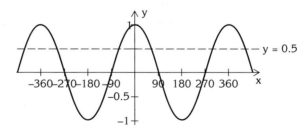

It is also possible to solve quadratic trigonometric equations, such as:

$$\tan^2 x + 5\tan x + 6 = 0$$

by factorizing the quadratic to give:

$$(\tan x + 2)(\tan x + 3) = 0$$

and then solving the equations tan $x = -2$ and tan $x = -3$. If the equation does not factorize easily, it may be necessary to use the quadratic equation formula.

Example 1:

Solve the equation $4 \sin x = 3$, giving all the solutions in the range $0 \leq x \leq 360°$.

Solution:

First solve for sin x, to give:

$$\sin x = \frac{3}{4}$$

Using a calculator gives $x = 48.6°$ to one decimal place. The graph below shows that there are two solutions in the range $0 \leq x \leq 360°$.

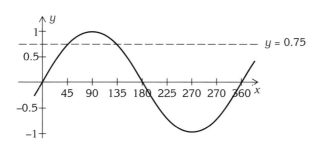

The second solution is given by:

$$x = 180 - 48.6 = 131.4°$$

Example 2:

Solve the equation $5 = 6\cos^2 \theta + \sin \theta$, giving all the solutions in the range $0 \leq \theta \leq 360°$.

Solution:

Using the identity $\cos^2 \theta + \sin^2 \theta = 1$, the $\cos^2 \theta$ term can be eliminated to give:

$$5 = 6\cos^2 \theta + \sin \theta$$
$$5 = 6(1 - \sin^2 \theta) + \sin \theta$$
$$6 \sin^2 \theta - \sin \theta - 1 = 0$$

This is a quadratic in sin θ and can be solved by factorizing:

$$6 \sin^2 \theta - \sin \theta - 1 = 0$$
$$(2\sin \theta - 1)(3\sin \theta + 1) = 0$$

So this equation has two solutions:

$$\sin \theta = \frac{1}{2} \quad \text{and} \quad \sin \theta = -\frac{1}{3}$$

The graph on page 218 illustrates the corresponding values for θ.

217

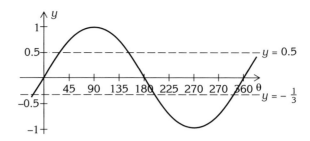

For $\sin\theta = \dfrac{1}{2}$ the solutions are $\theta = 30°$ and $\theta = 180 - 30 = 150°$.

For $\sin\theta = -\dfrac{1}{3}$ a calculator will give $\theta = -19.5°$, which is not in the required range

of values.

The solutions in the range $0 \le \theta \le 360°$ are $\theta = 180 + 19.5 = 199.5°$
and $\theta = 360 - 19.5 = 340.5°$.

source: the starting point of a *directed network*.

spanning tree: a subgraph that includes all the *vertices* in the original *graph* and is also a tree.

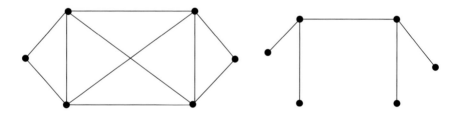

Spearman's rank correlation coefficient: given by

$$r_s = 1 - \frac{6\Sigma d^2}{n(n^2 - 1)}$$

It is used to calculate the product moment *correlation* coefficient between ranked bivariate data. The term d is the rank difference for each pair of values.

For tied ranks, each place is given the average rank of the equal values. For example, if the top two items are considered equal then they would be both given a rank of 1.5, the mean of 1 and 2.

Example:

Two judges at a flower show placed entries in the sweet pea category in the order shown in the table on page 219. Calculate Spearman's ranked correlation coefficient for this data.

Entry	A	B	C	D	E	F	G
Judge X	4	3	6	2	7	5	1
Judge Y	4	2	6	1	5	7	3

Solution:

Entry	A	B	C	D	E	F	G
Judge X	4	3	6	2	7	5	1
Judge Y	4	2	6	1	5	7	3
d	0	1	0	1	2	−2	−2

Here $\Sigma d^2 = 14$ and Spearman's ranked correlation coefficient is

$$r_s = 1 - \frac{6\Sigma d^2}{n(n^2 - 1)} = 1 - \frac{6 \times 14}{7(49 - 1)} = 1 - 0.25 = 0.75$$

This indicates some positive correlation between the two judges.

speed: the magnitude of the *velocity vector*. Speed is a *scalar quantity*.

For an object moving in a straight line at constant speed, the relationship between speed, distance and time is:

speed $=$ distance $\times$ time

sphere: the volume of a sphere of radius r is $\dfrac{4\pi r^3}{3}$ and its surface area is $4\pi r^2$.

square root: the square root of a number, n say, is a number whose square equals the number itself; i.e. if m is the square root of n then $m^2 = n$. It is written as $\sqrt{n}$ or $n^{1/2}$.

For example the square root of 9 is −3 or +3.

standard deviation: the standard deviation is the square root of the *variance*. The standard deviation of a population is denoted by σ and the standard deviation of a sample is denoted by s and can be found using the formula:

$$s = \sqrt{\frac{\Sigma(x - \bar{x})^2}{n}} = \sqrt{\left(\frac{\Sigma x^2}{n} - \bar{x}^2\right)}$$

The second of these is often easier to use.

Most calculators will calculate both the population and sample standard deviations.

standard error (of the mean): the *standard deviation* of a distribution of means around their mean is known as the standard error of the mean. The standard error of the sample means, $\sigma_{\bar{x}}$ is given by $\sigma/\sqrt{n}$, where n is the sample size.

Example:

A fair die is thrown four times and the mean score calculated. This is repeated 30 times. The results are shown below:

5.25, 4, 3.75, 4, 3.75, 2.5, 3, 3.5, 2.75, 2, 4, 2.25, 4.25, 4, 3.75, 3.5, 3.75, 4, 4.75, 5, 4.25, 2, 3.75, 2.75, 3.5, 3, 4.25, 3.5, 3.25, 3.75.

Calculate the standard error of the means.

Solution:

For this data the standard deviation is 0.789 and the sample size $n = 4$. Therefore the standard error of the means is

$$\frac{0.789}{\sqrt{4}} = 0.3945$$

standard form: in standard form, numbers are written in the form $a \times 10^n$, where $1 \leq a < 10$ and n is an integer. For example $4\,320\,000 = 4.32 \times 10^6$, and $0.00000451 = 4.51 \times 10^{-6}$.

standard normal distribution: see *normal distribution*.

star graph: a complete *bipartite* graph in which the number of *vertices* in the first subset is 1.

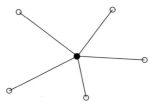

static equilibrium: see *equilibrium*.

stationary points: a stationary point or turning point is a point on a curve where its gradient is zero. Stationary points can be classified as either *local maxima*, *local minima* or *points of inflexion*. The graph below shows the curve $y = x^5 - 2x^3$, which has three stationary points.

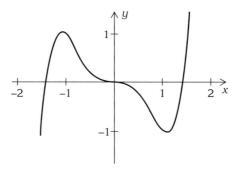

There is a local maximum at $x = -1$, a point of inflexion at $x = 0$ and a local minimum at $x = 1$.

To find the stationary points of a curve solve the equation $dy/dx = 0$. This will indicate the positions of the stationary points. To find the type of stationary point, either consider the value of dy/dx on either side of the stationary point or use the value of the second derivative. At a stationary point:

if $\dfrac{d^2y}{dx^2} > 0$ then it is a local minimum

if $\dfrac{d^2y}{dx^2} < 0$, then it is a local maximum

if $\dfrac{d^2y}{dx^2} = 0$, then the second derivative cannot be used to determine the type of stationary point and a value of $\dfrac{dy}{dx}$ on either side will have to be considered

Example:

Find and classify the turning points of $y = x^5 - 4x^4 + 4x^3$.

Solution:

First differentiate to obtain $\dfrac{dy}{dx} = 5x^4 - 16x^3 + 12x^2$.

For the stationary points, $\dfrac{dy}{dx} = 0$, so

$$5x^4 - 16x^3 + 12x^2 = 0$$
$$x^2(5x^2 - 16x + 12) = 0$$
$$x^2(5x - 6)(x - 2) = 0$$

$$x = 0 \quad \text{or} \quad x = \frac{6}{5} \quad \text{or} \quad x = 2$$

This shows that there are three stationary points at $x = 0$, $x = 1.2$ and $x = 2$. Now the second derivative can be used to determine the nature of each stationary point.

$$\frac{d^2y}{dx^2} = 20x^3 - 48x^2 + 24x$$

Substituting $x = 1.2$ gives

$$\frac{d^2y}{dx^2} = 20 \times 1.2^3 - 48 \times 1.2^2 + 24 \times 1.2$$

$$= -5.76$$

As this is negative there is a local maximum at $x = 1.2$, with coordinates (1.2, 1.10592).

Substituting $x = 2$ gives

$$\frac{d^2y}{dx^2} = 20 \times 2^3 - 48 \times 2^2 + 24 \times 2$$

$$= 16$$

As this is positive there is a local minimum at $x = 2$, with coordinates (2, 0).

For $x = 0$ the second derivative will clearly be zero as x is a factor of every term, so the first derivative must be considered for values of x close to 0, as shown below.

$x = -1$	$x = 0$	$x = 1$
$\dfrac{dy}{dx} = 33$	$\dfrac{dy}{dx} = 0$	$\dfrac{dy}{dx} = 1$

As the gradient is positive on both sides of the turning point, there is a point of inflexion at $x = 0$, with coordinates (0,0).

stem and leaf diagram: a method of presenting data, using the items of data themselves.

The stem consists of one or more of the leading digits and the leaf consists of the remaining digit. The stem values are listed vertically in ascending order and the leaves added to the appropriate stem also in ascending order.

Each stem and leaf diagram should have the units of the leaves clearly displayed.

The *median* can readily be obtained from a stem and leaf diagram and any *skewness* can be easily observed.

Example 1:

Construct a stem and leaf diagram for the following set of data. Locate the median and comment on the skewness of the distribution.

64 75 37 48 59 50 68 65 43 52 60 79 76 83 56 88 72 65 63 79

Solution:

Unit is 1

```
3 | 7
4 | 3 8
5 | 0 2 6 9
6 | 0 3 4 5 5 8
7 | 2 5 6 9 9
8 | 3 8
```

The median is the middle value, which is the mean of the 10th and 11th values, which are 64 and 65. Therefore the median is 64.5. The data is slightly negatively skewed.

Example 2:

Construct a stem and leaf display of the following data.

10 310 190 70 150 20 190 140 220 200 490 230 10 340 190

Solution:

Unit = 10

```
0 | 1 1 2 7
1 | 4 5 9 9 9
2 | 0 2 3
3 | 1 4
4 | 9
```

stiffness: a perfectly elastic spring (or elastic string) is characterized by two positive constants: its *natural length* and its *stiffness*. *Hooke's Law* provides a linear model relating the tension in the spring or string:

tension = stiffness × extension

The stiffness constant is sometimes written in terms of a quantity called the modulus of elasticity, where:

$$\text{stiffness} = \frac{\text{modulus of elasticity}}{\text{natural length}}$$

stochastic model: a model that allows for an element of chance in the outcome.

stratified random sampling: in stratified random sampling, a *population* is divided into distinct sections or strata. A *sample* is taken randomly from each stratum. The size of the sample taken from each stratum is in proportion to the size of that stratum in the population.

For example, a survey of 50 is required from a population of 356 people of whom 197 are men. The population can be divided into two strata: male and female. The proportions of each that should be sampled are:

Men:$\qquad \dfrac{197}{356} \times 50 = 27.67$

Women:$\qquad \dfrac{159}{356} \times 50 = 22.33$

Therefore a stratified random sample of this population should consist of 28 men and 22 women.

Stratified random sampling has the advantage that it distributes the sample more evenly over the population. It has the disadvantage that it requires prior knowledge of the population.

Student t test: see *t test*.

subdivision of a graph: a *graph* formed by inserting *vertices* of degree 2 into a graph.

subgraph: part of a *graph* which is itself a graph.

substitution methods for integration: see *integration by direct substitution* and *integration using trigonometric substitutions*.

sufficient: a term used to describe a relationship between two statements. If P is a sufficient condition for Q, this means that whenever P is true, then Q is also true. We often say P implies Q or write $P \Rightarrow Q$ or $Q \Leftarrow P$. However it does not mean that if Q is true then P is also true, although this may be so in certain cases.

Consider the following example.

Let P be the statement $x > 4$.

Let Q be the statement $x^2 > 16$.

We say that P is a sufficient condition for Q, because, whenever P is true, then Q is also true. But note that when Q is true, then P may not be true, for example if $x = -5$.

See *necessary* and *necessary and sufficient*.

sum and product of the roots of a quadratic equation: for the quadratic equation $ax^2 + bx + c = 0$ it is possible to find simple formulas for the sum and the product of the roots of the equation. If α and β are the roots of the quadratic equation, then:

$$(x - \alpha)(x - \beta) = 0$$
$$x^2 - (\alpha + \beta)x + \alpha\beta = 0$$

The general form of the quadratic equation can be expressed as:

$$x^2 + \frac{b}{a}x + \frac{c}{a} = 0$$

Comparing these two forms of the quadratic equation gives:

$$\alpha + \beta = -\frac{b}{a} \qquad \text{an expression for the sum of the roots}$$

and

$$\alpha\beta = \frac{c}{a} \qquad \text{an expression for the product of the roots}$$

The sum and difference are often used to find a new quadratic equation that has roots related to the roots of the original. This is illustrated in the example below.

Example:

The quadratic equation $x^2 - 7x + 2 = 0$ has roots α and β. Find the equations that have roots:

(a) $\alpha + 1$ and $\beta + 1$

(b) $\dfrac{1}{\alpha^2}$ and $\dfrac{1}{\beta^2}$

Solution:

First note that, in the quadratic equation, $a = 1$, $b = -7$ and $c = 2$.

Then, using $\alpha + \beta = -\dfrac{b}{a}$ and $\alpha\beta = \dfrac{c}{a}$

gives

$$\alpha + \beta = -\frac{(-7)}{1} = 7$$

and

$$\alpha\beta = \frac{2}{1} = 2$$

(a) For the equation with roots $\alpha + 1$ and $\beta + 1$:

$$
\begin{aligned}
\text{Sum of new roots} &= (\alpha + 1) + (\beta + 1) \\
&= (\alpha + \beta) + 2 \\
&= 7 + 2 \\
&= 9
\end{aligned}
$$

$$
\begin{aligned}
\text{Product of new roots} &= (\alpha + 1) \times (\beta + 1) \\
&= \alpha\beta + \alpha + \beta + 1 \\
&= \alpha\beta + (\alpha + \beta) + 1 \\
&= 2 + 7 + 1 \\
&= 10
\end{aligned}
$$

So, for the new equation,

$$-\frac{b}{a} = 9 \qquad \text{and} \qquad \frac{c}{a} = 10$$

If $a = 1$, then $b = -9$ and $c = 10$, giving the equation $x^2 - 9x + 10 = 0$.

(b) For the equation with roots, $\dfrac{1}{\alpha^2}$ and $\dfrac{1}{\beta^2}$

$$\text{Sum of new roots} = \frac{1}{\alpha^2} + \frac{1}{\beta^2}$$

$$= \frac{\alpha^2 + \beta^2}{\alpha^2\beta^2}$$

$$= \frac{(\alpha + \beta)^2 - 2\alpha\beta}{(\alpha\beta)^2}$$

$$= \frac{7^2 - 2 \times 2}{2^2}$$

$$= \frac{45}{4}$$

$$\text{Product of new roots} = \frac{1}{\alpha^2} \times \frac{1}{\beta^2}$$

$$= \frac{1}{(\alpha\beta)^2}$$

$$= \frac{1}{2^2}$$

$$= \frac{1}{4}$$

So, for the new equation, $-\dfrac{b}{a} = \dfrac{45}{4}$ and $\dfrac{c}{a} = \dfrac{1}{4}$.

If $a = 4$, then $b = -45$ and $c = 1$, giving the equation $4x^2 - 45x + 1 = 0$.

supersink: an artificial *sink* introduced into a *network* and fed from all the sinks.

supersource: an artificial *source* introduced into a *network* and feeding all the sources.

surds: a surd is an expression that contains one or more irrational roots (see *irrational numbers*), such as $\sqrt{2}$ or $\sqrt{3}$. It is usual to express surds in their simplest forms, for example $\sqrt{28}$ would be written as $2\sqrt{7}$ since $\sqrt{28} = \sqrt{(4 \times 7)}\ = \sqrt{4} \times \sqrt{7} = 2\sqrt{7}$.

There are several rules that apply when working with surds, that are expressed in their simplest forms.

Multiplication	$a\sqrt{b} \times c\sqrt{d} = ac\sqrt{bd}$
Division	$\dfrac{a\sqrt{b}}{c\sqrt{d}} = \dfrac{a}{c}\sqrt{\left(\dfrac{b}{d}\right)}$
Addition	$a\sqrt{b} + c\sqrt{b} = (a + c)\sqrt{b}$
	$a\sqrt{b} + c\sqrt{d} \quad$ no simplification possible
Subtraction	$a\sqrt{b} - c\sqrt{b} = (a - c)\sqrt{b}$
	$a\sqrt{b} - c\sqrt{d} \quad$ no simplification possible

Example:

Write the following surds in their simplest forms.

(a) $\sqrt{48}$

(b) $3\sqrt{2} \times 5\sqrt{6}$

(c) $5\sqrt{12} + 7\sqrt{3}$

(d) $\dfrac{3\sqrt{42}}{5\sqrt{7}}$

Solution:

(a) $\sqrt{48} = \sqrt{16 \times 3}$

$\quad\quad\quad = \sqrt{16} \times \sqrt{3}$

$\quad\quad\quad = 4\sqrt{3}$

(b) $3\sqrt{2} \times 5\sqrt{6} = 3 \times 5 \sqrt{2 \times 6}$

$\quad\quad\quad\quad\quad\quad = 15\sqrt{12}$

$\quad\quad\quad\quad\quad\quad = 15 \sqrt{(4 \times 3)}$

$\quad\quad\quad\quad\quad\quad = 15(\sqrt{4} \times \sqrt{3})$

$\quad\quad\quad\quad\quad\quad = 30\sqrt{3}$

(c) $5\sqrt{12} + 7\sqrt{3} = 5 \sqrt{(4 \times 3)} + 7\sqrt{3}$

$\quad\quad\quad\quad\quad\quad = 5\sqrt{4} \sqrt{3} + 7\sqrt{3}$

$\quad\quad\quad\quad\quad\quad = 10\sqrt{3} + 7\sqrt{3}$

$\quad\quad\quad\quad\quad\quad = 17\sqrt{3}$

(d) $\dfrac{3\sqrt{42}}{5\sqrt{7}} = \dfrac{3}{5} \times \sqrt{\dfrac{42}{7}}$

$\quad\quad\quad\quad = \dfrac{3}{5} \times \sqrt{6}$

$\quad\quad\quad\quad = \dfrac{3\sqrt{6}}{5}$

(See also *rationalization of denominators*.)

surface forces: see *contact forces*.

symmetric data: a distribution is symmetrical if the difference between the mean and the median (mean–median) is zero. An appropriate *pictorial representation* of the data, (*histogram*, *stem and leaf diagram*, etc.) would produce a mirror image about the center.

(See also *skewness*.)

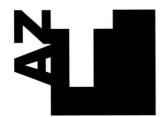

t distribution: a *continuous* *random variable* X which has the *probability density function*

$$f(x) = T_v\left(1 + \frac{x^2}{v}\right)^{-\frac{v+1}{2}}$$

for $-\infty < x < \infty$ where T_v is a constant is said to have a t distribution.

The parameter v is called the *degrees of freedom* and for a particular value of v, the appropriate t distribution is denoted by T_v. Values of the t distribution are tabulated for various degrees of freedom and the t distribution is used in the "Student t test" for hypothesis testing.

t test (Student t test): in *hypothesis testing*, the t test is used to test for differences between means when small samples are involved ($n \leq 30$ say). For larger samples use the *z test*.

The t test can test:

- if a sample has been drawn from a *normal population* with known mean and variance (single sample)

- if two unknown population means are identical given two independent random samples (two unpaired samples)

- if two paired random samples come from the same normal population (two paired samples [paired differences]).

Single sample test:

Let $x_1, x_2, \ldots, x_n$ be a random sample with mean $\bar{x}$ and variance s^2. To test if this sample comes from a normal population with known mean μ and unknown variance σ^2, the test statistic

$$T = \frac{\bar{x} - \mu}{s/\sqrt{n}}$$

is used to test the null hypothesis H_0: the population mean equals μ. If the test statistic lies in the critical region, the critical values of which are found from the distribution of $T_{v,\alpha}$, H_0 is rejected in favor of the alternative hypothesis H_1. v are the degrees of freedom; for a single sample test $v = n - 1$, and α is the significance level of the test.

Example:

It is claimed that the length of a certain species of snake is normally distributed with mean 44 cm. A sample of 21 such snakes was taken. The mean length of the sample was found to be 42 cm and the sample variance was calculated to be 36 cm². Is there any evidence at the 5% level of significance against the claim that the population mean is 44 cm?

Solution:

Here $\mu = 44$, $\bar{x} = 42$, $s = \sqrt{36} = 6$ and $n = 21$

This is a "two-tailed test" (see *alternative hypothesis*) since we are looking for any difference.

$H_0: \mu = 44$

$H_1: \mu \neq 44$

The test statistic is given by

$$T = \frac{42 - 44}{6/\sqrt{21}}$$

$$= -1.528.$$

To find the critical values we need $T_{20,\,0.025}$, since $v - 1 = 20$ and it is a two-tailed test. From tables $T_{20,\,0.025} = 2.086$

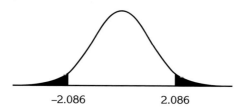

| −2.086 | 2.086 |

Since our test statistic does not fall in the critical region, there is insufficient evidence to reject H_0. We therefore conclude that the population mean length of snake is 44 cm.

When you are using a graphic calculator to perform this hypothesis test, the p value for $T = -1.528$ is $p = 0.1423$, which is greater than the significance level of 0.025. This leads to the conclusion that we cannot reject H_0.

Two unpaired samples:

Let $x_1, x_2, \ldots, x_m$ be a random sample with mean $\bar{x}$ and variance s_x^2 drawn from a normal population with unknown mean μ_x and unknown variance σ_x^2.

Let $y_1, y_2, \ldots, y_n$ be a random sample with mean $\bar{y}$ and variance s_y^2 drawn from a normal population with unknown mean μ_y and unknown variance σ_y^2.

To test the null hypothesis that the two unknown population means are the same we use the test statistic

$$T = -\frac{\bar{x} - \bar{y}}{\hat{\sigma}\sqrt{(1/m + 1/n)}}$$

where the estimate of the common population standard deviation, $\hat{\sigma}$

$$\hat{\sigma} = \sqrt{\left(\frac{ms_x^2 + ns_y^2}{(m-1) + (n-1)}\right)}$$

The test statistic T is distributed T_v, where $v = (m - 1) + (n - 1)$ for two unpaired samples. If the test statistic lies in the critical region whose critical values are found from the distribution of $T_{v,\,\alpha}$, H_0 is rejected in favor of the alternative hypothesis H_1.

Example:

A researcher investigating the effects of pollution on two rivers takes an independent random sample of fish of a certain species from each river, measures their mass in ounces and obtains the following results.

River 1	20	10	17	7	10	18			
River 2	14	6	10	8	9	7	7	6	8

Test at the 5% level of significance if there is any evidence of a difference in the mean weight of fish between the two rivers.

Solution:

Assume that each sample is taken from a normal population.

Here $m = 6$, $\bar{x} = 13.667$, $s_x^2 = 23.556$

$n = 9$, $\bar{y} = 8.333$, $s_y^2 = 5.556$

The estimate of the standard deviation is

$$\hat{\sigma} = \sqrt{\left(\frac{ms_x^2 + ns_y^2}{(m-1) + (n-1)}\right)} = \sqrt{14.72} = 3.84$$

Let μ_1 be the population mean of river 1, and μ_2 be the population mean of river 2.

$H_0: \mu_1 = \mu_2$

$H_1: \mu_1 \neq \mu_2$

The test statistic is

$$T = \frac{13.667 - 8.333}{3.84\sqrt{(\frac{1}{6} + \frac{1}{9})}} = 2.64$$

To find the critical values we need $T_{13,\,0.025}$, since $v = (6-1)+(9-1) = 13$ and it is a two-tailed test.

From tables, $T_{13,\,0.025} = 2.1604$

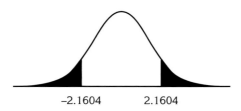

−2.1604   2.1604

Since our test statistic lies in the critical region we reject H_0 in favor of H_1. Therefore there is sufficient evidence at the 5% level to conclude that the mean weight of fish in River 1 is not equal to the mean weight of fish in River 2.

When using a graphic calculator to perform this hypothesis, test the *p value* for $T = 2.64$ is 0.0204, which is less than the significance level of 0.025. This leads to the conclusion that we can reject H_0.

t test (Student t test)

Two paired samples:

Let (X_1, Y_1), (X_2, Y_2), ... , (X_n, Y_n) be n independent pairs of values and $D_1 = X_1 - Y_1$, $D_2 = X_2 - Y_2$, ... $D_n = X_n - Y_n$. The test statistic

$$T = \frac{\bar{D}}{s/\sqrt{n}}$$

is used to test the null hypothesis that both samples come from the same normal distribution. $\bar{D}$ and s are the mean and standard deviation of the paired sample differences, D_1, D_2 ... D_n. The test statistic T is distributed T_v, where $v = n - 1$ for two paired samples. If the test statistic lies in the critical region whose critical values are found from the distribution of $T_{v, \alpha}$, H_0 is rejected in favor of the alternative hypothesis H_1.

Example:

A drug G, which reduces high blood pressure, was administered to 12 patients. In a later trial the same quantity of a drug PH, which also reduces high blood pressure was given to the same 12 patients. The table below gives the concentration of each drug in the bloodstream one hour after administration.

Patient	1	2	3	4	5	6	7	8	9	10	11	12
G	2.8	5.8	4.3	3.8	3.5	4.0	5.6	4.1	3.5	2.9	3.0	2.9
PH	3.1	6.7	4.2	3.6	2.9	4.4	5.7	4.5	3.4	2.9	4.3	3.5

The experimenter predicted (before the experiment) that the concentration of PH would be greater than the concentration of G on average. Is there any evidence to support this?

Solution:

Assume that the patients were chosen at random and that drug concentrations are normally distributed.

Let D be the difference between the concentration of G and the concentration of PH, i.e. concentration of G – concentration of PH. We wish to test for evidence that D is less than 0.

$$H_0 : \bar{D} \geq 0$$
$$H_1 : \bar{D} < 0$$

Patient	1	2	3	4	5	6	7	8	9	10	11	12
G	2.8	5.8	4.3	3.8	3.5	4.0	5.6	4.1	3.5	2.9	3.0	2.9
PH	3.1	6.7	4.2	3.6	2.9	4.4	5.7	4.5	3.4	2.9	4.3	3.5
Difference D	−0.3	-0.9	0.1	0.2	0.6	−0.4	−0.1	−0.4	0.1	0	−1.3	-0.6

For D we find that $\bar{D} = -0.25$ and $s = 0.518$.

Therefore the test statistic is

$$T = \frac{-0.25}{0.518/\sqrt{12}} = -1.672$$

We will test at the 5% significance level. To find the critical values we need $T_{11, 0.05}$, since $v = 12 - 1 = 11$ and it is a one-tailed test.

From tables, $T_{11,\,0.05} = 1.7959$. The critical value is -1.7959, since the alternative hypothesis is that $\bar{D} < 0$

-1.7959

The test statistic does not lie in the critical region so there is insufficient evidence to reject H_0. Therefore we conclude that the mean concentration of PH is not greater than the mean concentration for G.

When you are using a graphic calculator for this hypothesis test, the *p value* associated with $T = -1.672$ is 0.0613, which is greater than the significance level of 0.05. This leads to the conclusion that we cannot reject H_0.

tangent: consider triangle ABC in which the right angle is at B. The side opposite the angle at A is the opposite side and has length o and the side AB is the adjacent side and has length a. The tangent of the angle θ is defined as

$$\tan \theta \; = \; \frac{o}{a}$$

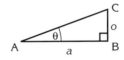

The tangent of an angle greater than 90° is also defined. The graph below shows the tangent function for angles between 0 and 720°.

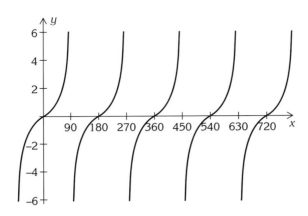

Note that the tangent function has a period of 180° (or π radians) and that the graph has vertical asymptotes at 90°, 270°, 450°, 630°, ...

tan⁻¹: this is the inverse of the *tangent* function. It is used when finding an angle given its *tangent*, and is often used in the context of right-angled triangles. When using a calculator you will be given the *principal value*, which will be between −90° and 90° ($-\pi/2$ and $\pi/2$ in radians). The graphs below show $y = \tan^{-1} x$ working in degrees on the first graph and radians on the second graph. Note that the function is defined for all values of x.

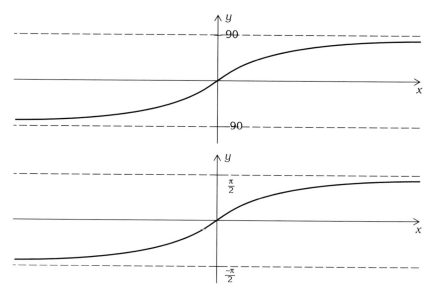

The derivative of $\tan^{-1}$ can be obtained by using the *chain rule* and a *trigonometric identity* in a similar way to the derivative of $\cos^{-1} x$. If $y = \tan^{-1} x$, then

$$\frac{dy}{dx} = \frac{1}{1 + x^2}$$

tangent to a curve: this is a straight line that touches the curve at a point and has the same gradient as the curve at that point. The diagram below shows three different tangents to the curve $y = x^3 - 3x^2$.

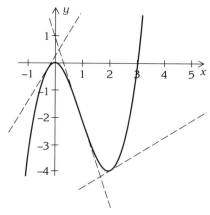

The equation of a tangent to a curve can be found by using the derivative to find its gradient.

Example:

Find the equation of the tangent to the curve $y = x^3 - 3x^2$, that passes through the point with coordinates $(1, -2)$.

Solution:

First find $\dfrac{dy}{dx}$ by differentiating y. This gives:

$$\frac{dy}{dx} = 3x^2 - 6x$$

When $x = 1$, $\dfrac{dy}{dx} = 3 \times 1 - 6 \times 1 = -3$

So the gradient of the tangent is -3. As the tangent is a straight line its equation is of the form $y = mx + c$, where m is the gradient, in this case -3. So the equation is $y = -3x + c$, where the value of c still has to be determined.
As the tangent passes through the point with coordinates $(1, -2)$, the values $x = 1$ and $y = -2$ can be substituted into $y = -3x + c$ to give $-2 = -3 + c$, or $c = 1$.
Then the equation of the tangent is $y = -3x + 1$. This is the middle tangent shown on the diagram on page 232.

tangent field: see *direction field*.

tension: the force experienced by an object attached to the end of a string, spring or rod when the string, spring or rod is being pulled along or stretched. A tension is often described as a "pulling force."

terminal velocity and terminal speed: when the motion of an object near to the Earth's surface is opposed by resistive forces (such as air resistance) which depend on the velocity of the object, the force of gravity and the resistive forces can come into balance. The velocity when this occurs is called the *terminal velocity*. The *terminal speed* is the magnitude of the terminal velocity. When the terminal speed has been reached the object will continue with this speed.

Example:

A parachute provides a parachutist with a resistance force of $20v^2$. After jumping from a hot-air balloon, a parachutist of mass 80 kg travels in a straight line. Find the terminal speed of the parachutist.

Describe the motion of the parachutist.

Solution:

The parachutist reaches the terminal speed when $80g = 20v^2$ (see the diagram on the next page). Solving for v we get

$$\text{terminal speed} = \sqrt{\frac{80g}{20}} = 6.26 \text{ m s}^{-1}$$

The parachutist jumps from the balloon and will initially fall under the forces of gravity and air resistance (small).

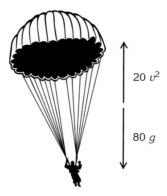

$20\ v^2$

$80\ g$

When the parachute opens, the parachutist will either continue to accelerate to a speed of 6.26 m s^{-1} if the speed on opening the parachute is less than 6.26 m s^{-1} or will decelerate to the terminal speed 6.26 m s^{-1} if the speed is already greater than 6.26 m s^{-1}. Subsequently, the parachutist will fall with a constant speed of 6.26 m s^{-1}.

test statistic: in *hypothesis testing*, the test statistic is a statistic (number) on which the decision to accept or reject the *null hypothesis* is based. For example, in the *chi squared* (χ^2) *test*, if the calculated test statistic is greater than the *critical value* of the test, then the null hypothesis is rejected in favor of the *alternative hypothesis*.

tetrahedron: a tetrahedron is a solid with four triangular faces.

thrust: the force experienced by an object attached to the end of a (compressed) spring or rod when the spring or rod is being pushed or compressed. A thrust is often described as a "pushing force."

time series: A time series is any series of data observed and recorded graphically at regular intervals of time. Time series analysis distinguishes between the general trend, seasonal variations (recurring fluctuations due to the seasons) and cyclic variations (fluctuations in the general trend). *Moving averages* of suitable order will give the general trend.

tonne: often referred to as a metric ton, it is a unit of mass equal to 1000 kg.

toppling: consider a box placed on a horizontal surface as in the diagram below.

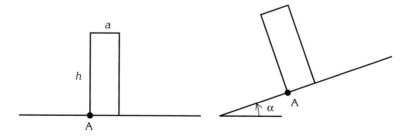

Suppose that the surface is gradually tipped so that the angle to the horizontal, α, increases. One of two things will happen to the box:

- it could slide down the slope
- it could topple over about corner A

Which of these happens depends on the coefficient of friction between the box and the surface and the aspect ratio of the box a/h.

The forces on the box are the force of gravity, mg, the normal reaction R and the friction F.

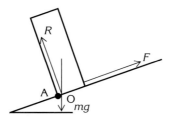

The three forces act through a common point O. Let OA = x.

Resolving forces up and perpendicular to the slope:

$R = mg \cos \alpha$ and $F = mg \sin \alpha$

Take moments of the forces about the corner A:

$mg \cos \alpha \times (a/2) = Rx + mg \sin \alpha \times (h/2)$

Solving for x we get

$$x = \frac{a \cos \alpha - h \sin \alpha}{2 \cos \alpha}$$

The box will topple before it slides if $F < \mu R$ and $x = 0$. This occurs when

$\mu > \tan \alpha$ and $\tan \alpha = a/h$

where μ is the coefficient of friction between the two surfaces.

torque: of a force or a couple is the moment of the force or couple.

(See *moment of a force*.)

tour: a route that visits every *vertex* in a network.

tour improvement algorithm: once the *nearest neighbor algorithm* has been applied to a *traveling salesperson algorithm*, it can be improved by applying this *algorithm*.

    Step 1: put your solution in a list: $V_1, V_2, \ldots V_n$.

    Step 2: let $i = 1$

    Step 3: if $d(V_i, V_{i+2}) + d(V_{i+1}, V_{i+3}) < d(V_i, V_{i+1}) + d(V_{i+2}, V_{i+3})$, then swap V_{i+1} and V_{i+2}.

    Step 4: replace i by $i + 1$.

    Step 5: if $i \leq n$, then go to step 2.

Once an improvement has been made, the algorithm need not be applied any more. $d(V_i, V_{i+2})$ is the distance between vertices V_i and V_{i+2}.

transformations of functions, include stretches, translations and reflections. The graph of $y = f(x)$ can be transformed to give the graphs of $y = f(x) + a$, $y = f(x + a)$, $y = af(x)$ and $y = f(ax)$, where a is a constant. The graphs on the next page illustrate these transformations for the case $f(x) = x^2$.

$y = f(x) + a$ produces a vertical translation of a units of the graph of $y = f(x)$.

$f(x) = x^2$

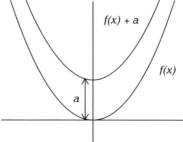

$y = f(x + a)$ produces a horizontal translation of a units of the graph of $y = f(x)$. Note that, when a is positive, the translation is to the left.

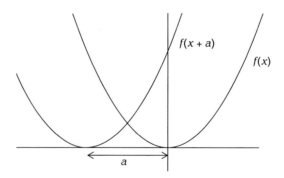

$y = af(x)$ produces a vertical stretch by a factor of a of the graph of $y = f(x)$. Note that, when a is negative, the graph is stretched and reflected in the x axis.

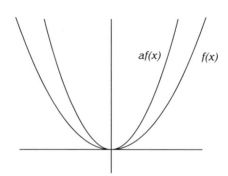

$y = f(ax)$ produces a horizontal stretch by a factor of $1/a$ of the graph of $y = f(x)$. Note that, when a is negative, the graph is stretched and reflected in the y axis.

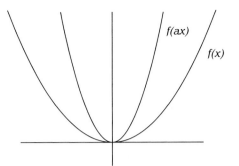

translation: a translation moves an object so that its shape and orientation do not change in any way. A *vector* is usually used to describe a translation.

The graph of $y = (x - 2)^2 - 3$ can be obtained by translating the graph of $y = x^2$ by the vector $\begin{pmatrix} 2 \\ -3 \end{pmatrix}$ as shown below.

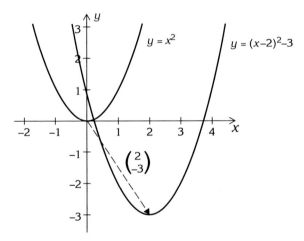

transposition of formulas: the process of rearranging a formula so that a specified variable becomes the subject of the formula. For example the formula $v = u + at$ can be transposed to make u the subject to give:

$$u = v - at$$

or so that a becomes the subject:

$$a = \frac{v - u}{t}$$

or with t as the subject:

$$t = \frac{v - u}{a}$$

When rearranging a formula you must apply the same operation to both sides of the formula in just the same way as when solving equations.

Examples:

(a) Make x the subject of the formula $A = \dfrac{1}{2}(x + y)h$

(b) Make x the subject of the formula $y = \dfrac{x + 3}{x - 2}$

Solution:

(a) First multiply both sides by 2 to give:

$$2A = (x + y)h$$

Then divide both sides by h to give:

$$\frac{2A}{h} = x + y$$

Subtracting y from both sides gives the required result:

$$\frac{2A}{h} - y = x$$

or

$$x = \frac{2A}{h} - y$$

(b) First multiply both sides by $(x - 2)$ to give:

$$y(x - 2) = x + 3$$

Removing the brackets gives:

$$xy - 2y = x + 3$$

Then bring all the terms containing x to the left-hand side and all the other terms to the right-hand side, to give:

$$xy - x = 2y + 3$$

Factorizing the left-hand side gives:

$$x(y - 1) = 2y + 3$$

Finally, dividing by $(y - 1)$ produces the required result:

$$x = \frac{2y + 3}{y - 1}$$

trapezium rule: a numerical method for approximating the value of an integral. Interpret a definite integral as an area under a graph; then the trapezium rule is obtained by dividing the area into n trapeziums, each with width h. The area under the curve is the sum of the areas of the trapeziums.

The formula for the trapezium rule is:

$$\int_b^a f(x)\, dx \approx \frac{h}{2}\left[(y_0 + y_n) + 2(y_1 + y_2 + \dots y_{n-1})\right]$$

$h = \dfrac{b - a}{n}$ is called the step length

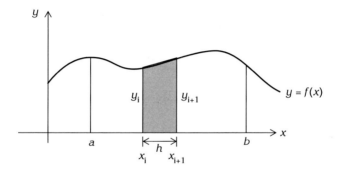

Example:

Find an approximate value of the integral $\int_0^1 e^{x^2}\,dx$ using the trapezium rule and five trapeziums.

Solution:

With an interval $(0,1)$ and five steps the width of each trapezium is $h = 0.2$. So the trapezium rule gives

$$\int_0^1 e^{x^2}\,dx \approx \frac{0.2}{2}[e^0 + e^1 + 2(e^{0.2^2} + e^{0.4^2} + e^{0.6^2} + e^{0.8^2})] = 1.48$$

For an improved value we would reduce the step length h.

traveling salesperson problems: require finding the shortest route through a *network* that visits each *vertex* only once. Although *nearest neighbor, lower bound* and *tour improvement* can be used to solve this type of problem, there is no simple *algorithm* for solving such problems (at the time of printing).

tree: a *connected graph* that contains no *cycles*.

tree diagram: a tree diagram is used to find probabilities of *combined events*. Probabilities for single *events* are written on the branches.

Branches from a point must cover all possible outcomes so that their probabilities add up to 1.

Each route along the branch leads to a combined event. To find the probability of the combined event multiply the probabilities along the branches.

Example 1:

Ten percent of the articles produced on an assembly line are defective in some way. An inspector takes two articles at random and examines them. What is the probability that:

(a) both are defective   (b) both are satisfactory   (c) at least one is defective?

Solution:

This information can be displayed on a tree diagram.
Let D be the event that the article is defective. Therefore $P(D) = 0.1$
Let $\bar{D}$ be the event that the article is not defective. Therefore $P(\bar{D}) = 0.9$

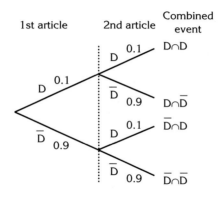

(a) The probability that they are both defective = P(D ∩ D) = 0.1 × 0.1 = 0.01
(b) The probability that they are both satisfactory = P(D̄ ∩ D̄) = 0.9 × 0.9 = 0.81
(c) The probability that at least one is defective = P(D ∩ D̄) + P(D̄ ∩ D)
$$= 0.1 × 0.9 + 0.9 × 0.1 = 0.18$$

Example 2:

A box contains 14 blue beads and 6 red beads. Two beads are picked at random without replacement. Use a tree diagram to find the probability that (a) both beads are blue and (b) they are different colors.

Solution:

Let B be the event the bead is blue and R be the event that the bead is red. A tree diagram of the problem is shown here.

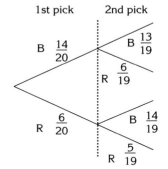

(a)   Probability both beads are blue is

$$\frac{14}{20} × \frac{13}{19} = \frac{182}{380} = \frac{91}{190} = 0.479$$

(b)   Probability that both beads are different colors is

$$\frac{14}{20} × \frac{6}{19} + \frac{6}{20} × \frac{14}{19} = \frac{168}{380} = \frac{42}{95} = 0.442$$

trial and improvement method: A trial and improvement method is used to solve equations, by substituting values of x into an equation of the form $f(x) = 0$, until a value of x is found that gives the solution to the required degree of accuracy. The solution will lie in the range $a < x < b$ if $f(a)$ and $f(b)$ have different signs.

Example:

Solve the equation $2x - \tan x = 0$.

Solution:

Substituting different values for x leads to the table below.

x	$2x - \tan x$	Comment
1	0.4426	
1.5	–11.1014	$1 < x < 1.5$
1.1	0.2352	$1.1 < x < 1.5$
1.2	–0.1722	$1.1 < x < 1.2$
1.16	0.0242	$1.16 < x < 1.2$
1.17	–0.0200	$1.16 < x < 1.17$
1.165	0.0025	$1.165 < x < 1.17$

The solution is then 1.17 to two decimal places.

triangle of forces: the rule used for adding forces. If two forces are represented in magnitude and direction by two sides of a triangle AB and BC then the third side of the triangle, AC, is the resultant of the two forces.

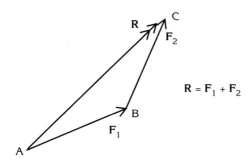

$$R = F_1 + F_2$$

triangle rule: a rule used for adding *vectors*.

trigonometry: the branch of mathematics that is concerned with the trigonometric functions. It includes, for example, finding lengths in triangles and the use of trigonometric functions in modeling.

trigonometric identities: the results that hold true for all angles. There are several standard identities which can be used to prove other results. These are the *Pythagorean identities*, the *trigonometric sum and difference formula*, the *factor formulas* and the *double angle formulas*.

Example:

Show that $\cos(3\theta) = 4\cos^3\theta - 3\cos\theta$

Solution:

First use the sum formula for cos $(A + B)$, with $A = 2\theta$ and $B = \theta$. This gives:

$$\cos(3\theta) = \cos(2\theta + \theta)$$
$$= \cos(2\theta)\cos\theta - \sin(2\theta)\sin\theta$$

Then, using the double angle formulas $\cos(2\theta) = 2\cos^2\theta - 1$ and $\sin(2\theta) = 2\sin\theta\cos\theta$ allows a further step:

$$\cos(3\theta) = \cos(2\theta)\cos\theta - \sin(2\theta)\sin\theta$$
$$= (2\cos^2\theta - 1)\cos\theta - 2\sin\theta\cos\theta\sin\theta$$
$$= 2\cos^3\theta - \cos\theta - 2\sin^2\theta\cos\theta$$

The $\sin^2\theta$ term can be eliminated using the Pythagorean identity $\sin^2\theta + \cos^2\theta = 1$ in the form $\sin^2\theta = 1 - \cos^2\theta$. This gives:

$$\cos(3\theta) = 2\cos^3\theta - \cos\theta - 2\sin^2\theta\cos\theta$$
$$= 2\cos^3\theta - \cos\theta - 2(1 - \cos^2\theta)\cos\theta$$
$$= 2\cos^3\theta - \cos\theta - 2\cos\theta + 2\cos^3\theta$$
$$= 4\cos^3\theta - 3\cos\theta$$

trigonometric ratios: for a right-angled triangle the trigonometric ratios are defined as:

$$\sin\theta = \frac{o}{h} \qquad \cos\theta = \frac{a}{h} \qquad \tan\theta = \frac{o}{a}$$

where the lengths are as shown in the triangle below.

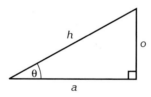

trigonometric sum and difference formulas: the sum and difference formulas for the trigonometric functions (sometimes called "compound angle formulas") are listed below:

$$\sin(A + B) = \sin A \cos B + \cos A \sin B$$
$$\sin(A - B) = \sin A \cos B - \cos A \sin B$$
$$\cos(A + B) = \cos A \cos B - \sin A \sin B$$
$$\cos(A - B) = \cos A \cos B + \sin A \sin B$$
$$\tan(A + B) = \frac{\tan A + \tan B}{1 - \tan A \tan B}$$
$$\tan(A - B) = \frac{\tan A - \tan B}{1 + \tan A \tan B}$$

Example:

Simplify

(a) $\sin(\theta + 180°)$
(b) $\cos(90° - \theta)$

Solution:

(a) Using the addition formula for sine gives:

$$\sin(\theta + 180°) = \sin\theta\cos 180° + \cos\theta\sin 180°$$

$$= \sin\theta \times (-1) + \cos\theta \times 0$$

$$= -\sin\theta$$

(b) Using the difference formula for cosine gives:

$$\cos(90° - \theta) = \cos 90°\cos\theta + \sin 90°\sin\theta$$

$$= 0 \times \cos\theta + 1 \times \sin\theta$$

$$= \sin\theta$$

The sum and difference formulas can be used in reverse to simplify expressions such as $a\sin\theta + b\cos\theta$.

If R and α are constants, then using the sine sum formula gives

$$R\sin(\theta + a) = R\cos\alpha\sin\theta + R\sin\alpha\cos\theta$$

$$= a\sin\theta + b\cos\theta$$

where $a = R\cos\alpha$ and $b = R\sin\alpha$

The values of R and α can be found

$$a^2 + b^2 = (R\cos\alpha)^2 + (R\sin\alpha)^2$$

$$= R^2(\cos^2\alpha + \sin^2\alpha)$$

$$= R^2$$

and

$$\frac{b}{a} = \frac{R\sin\alpha}{R\cos\alpha}$$

$$= \tan\alpha$$

Example:

Write $4\sin\theta + 7\cos\theta$ in the form $R\sin(\theta + \alpha)$.

Solution:

Using the results above

$$R^2 = 4^2 + 7^2 \qquad \text{and} \qquad \tan\theta = \frac{7}{4}$$

$$R = \sqrt{65} \qquad\qquad \theta = 60.3°$$

$$4\sin\theta + 7\cos\theta = \sqrt{65}\sin(\theta + 60.3)$$

trivial constraints: the limits on values of variables that give little information as to how they are related, such as $x \geq 0$, $y \geq 0$.

turning points: see *stationary points*.

two-tailed test: see *alternative hypothesis*.

type I error and type II error: see *errors*.

unbiased estimator: see *estimation*.

uniform acceleration: a problem in which the acceleration is said to be *uniform* is motion in a straight line with acceleration which is constant in magnitude. The *constant accelera-tion* formulas can be applied to such problems.

uniform body: an object with its mass distributed evenly thoughout its volume is called a *uniform body*. For such a body the density is constant and the *center of mass* is at the geometric center of the body (its *centroid*).

uniform circular motion: the motion of an object which moves in a circle, with constant speed. (See *circular motion*.)

uniform speed: see *constant speed*.

uniform velocity: a problem in which the velocity is said to be uniform is one that involves motion in a straight line, with constant speed. For such motion, acceleration is zero.

union: the union of two sets A and B is the set that contains all the elements found in either A or B.

For example, consider the two sets of numbers A = {1, 2, 4, 6, 9, 11, 15, 16} and B = {1, 3, 6, 7, 8, 14}, then the union of A and B is the set C = {1, 2, 3, 4, 6, 7, 8, 9, 11, 14, 15, 16}.

The notation for the union is $\cup$, so we write A $\cup$ B = C.

unit vector: a vector with magnitude equal to 1. For a given vector **a**, the vector

$$e = \frac{a}{|a|}$$

is a unit vector in the direction of the vector **a**.

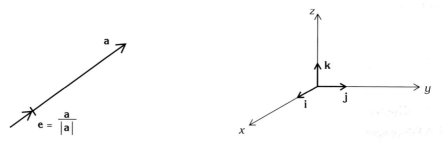

Unit vectors along the three-dimensional *Cartesian coordinate* system are denoted by **i**, **j** and **k** as shown in the second diagram above.

variance: the variance is the square of the standard deviation. For a population the variance is denoted by σ^2. The variance of a sample is a *measure of dispersion* for that sample. It is denoted by s^2 and is the mean squared deviation about the mean of the sample.

The variance of a sample of n values $x_1, x_2, ..., x_n$ is given by the formula:

$$s^2 = \frac{\Sigma(x-\bar{x})^2}{n}$$

This can be expressed as:

$$s^2 = \frac{\Sigma x^2}{n} - \bar{x}^2$$

for ease of calculation. It can be thought of as "the mean of the squares – the square of the mean."

For samples presented in frequency tables the variance is given by:

$$s^2 = \frac{\Sigma f(x-\bar{x})^2}{\Sigma f}$$

$$= \frac{\Sigma f x^2}{\Sigma f} - \bar{x}^2$$

For a group of frequency tables, the midpoint of each class is chosen as a representative value for x.

Example 1:

Calculate the mean and variance of the following sample

12, 15, 18, 14, 19, 16, 14, 13, 12, 15, 17, 18, 19, 15, 14, 13, 11, 15

Solution:

Here $n = 18$, $\Sigma x = 270$, $\Sigma x^2 = 4150$.

Therefore

$$\bar{x} = \frac{\Sigma x}{n} = \frac{270}{18} = 15$$

and

$$s^2 = \frac{\Sigma x^2}{n} - \bar{x}^2 = \frac{4150}{18} - 15^2 = 5.556 \quad \text{(to 3 d.p.)}$$

Example 2:

Find the mean and variance of the following sample.

Length x (cm)	150	151	152	153	154	155
Frequency f	2	3	5	4	2	1

Solution:

It is convenient to work in a table:

x	f	fx	fx^2
150	2	300	45000
151	3	453	68403
152	5	760	115520
153	4	612	93636
154	2	308	47432
155	1	155	24025
	$\Sigma f = 17$	$\Sigma fx = 2588$	$\Sigma fx^2 = 394\ 016$

Here

$$\bar{x} = \frac{\Sigma fx}{\Sigma f} = 152.235 \quad \text{(to 3 d.p.)}$$

and

$$s^2 = \frac{\Sigma fx^2}{\Sigma f} - \bar{x}^2 = \frac{394016}{17} - \left(\frac{2588}{17}\right)^2 = 1.827 \quad \text{(to 3 d.p.)}$$

vector: a vector is a mathematical object which has two properties:

- a magnitude or size
- a direction in space (or in the plane).

A vector is described pictorially by a "directed line segment," i.e. a line with an arrow attached.

It is important to note that the position of the vector is not part of the definition. Two vectors are equal if they have the same magnitude and direction, irrespective of where they are drawn.

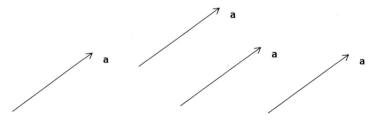

Vector addition is defined by a *parallelogram rule*:

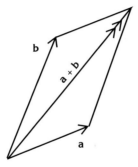

or alternatively by a *triangle rule*:

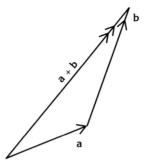

The sum of two vectors is often called the *resultant*.

The negative of a vector, −**a**, has the same magnitude as **a**, is parallel to **a** but points in the opposite direction to **a**.

Vector subtraction is defined by **a** − **b** = **a** + (−**b**).

Physical quantities that have magnitude and direction and add according to the parallelogram rule or the triangle rule can be modeled by vectors. Examples of such quantities are *force*, *displacement*, *velocity*, *acceleration*, *momentum* and *impulse*. (See also *components of a vector*.)

vector addition: see *vector*.

vector equation of a line: the vector equation of a line can be found in two cases:

- If the line passes through the point with position vector **a** and is parallel to the vector **b**, then the equation of the line is **r** = **a** + λ**b**.

- If the line passes through the points with position vectors **a** and **b**, then the equation of the line is **r** = **a** + λ(**b** − **a**).

Note that in the second example **b** – **a** is a vector parallel to the line.

Example:

Find the equation of the line that passes through the points with position vectors

$$\mathbf{a} = 3\mathbf{i} + 5\mathbf{j} - 12\mathbf{k} \quad \text{and} \quad \mathbf{b} = 4\mathbf{i} - 5\mathbf{j} + 6\mathbf{k}$$

Solution:

The vector $\mathbf{b} - \mathbf{a} = (4 - 3)\mathbf{i} + (-5 - 5)\mathbf{j} + (6 - (-12))\mathbf{k} = \mathbf{i} - 10\mathbf{j} + 18\mathbf{k}$.

So the vector equation of the line is

$$\mathbf{r} = 3\mathbf{i} + 5\mathbf{j} - 12\mathbf{k} + \lambda(\mathbf{i} - 10\mathbf{j} + 18\mathbf{k})$$

or, alternatively,

$$\mathbf{r} = 4\mathbf{i} - 5\mathbf{j} + 6\mathbf{k} + \lambda(\mathbf{i} - 10\mathbf{j} + 18\mathbf{k})$$

vector equation of a plane: if two vectors **a** and **b** are parallel to a plane that contains the point with position vector **c**, then the equation of the plane is $\mathbf{r} = \lambda\mathbf{a} + \mu\mathbf{b} + \mathbf{c}$.

If **n** is a vector that is perpendicular to the plane that contains the point with position vector **c**, then the vector equation of the plane is $\mathbf{r}.\mathbf{n} = \mathbf{c}.\mathbf{n}$.

Example:

A plane is perpendicular to the vector $\mathbf{i} + \mathbf{j} + \mathbf{k}$ and includes the point with position vector $2\mathbf{i} + 3\mathbf{j} + 4\mathbf{k}$. Find the vector equation of the plane and the Cartesian equation of the plane.

Solution:

The vector equation of the plane is given by $\mathbf{r}.\mathbf{n} = \mathbf{a}.\mathbf{n}$, which in this case gives:

$$\mathbf{r}.(\mathbf{i} + \mathbf{j} + \mathbf{k}) = (2\mathbf{i} + 3\mathbf{j} + 4\mathbf{k}).(\mathbf{i} + \mathbf{j} + \mathbf{k})$$
$$\mathbf{r}.(\mathbf{i} + \mathbf{j} + \mathbf{k}) = 2 + 3 + 4$$
$$\mathbf{r}.(\mathbf{i} + \mathbf{j} + \mathbf{k}) = 9$$

To find the Cartesian equation, let $\mathbf{r} = x\mathbf{i} + y\mathbf{j} + z\mathbf{k}$. Then

$$(x\mathbf{i} + y\mathbf{j} + z\mathbf{k}).(\mathbf{i} + \mathbf{j} + \mathbf{k}) = 9$$
$$x + y + z = 9$$

vector product: see *cross product*.

velocity: defined as the rate of change of displacement. It is a *vector* quantity, having both magnitude and direction. The magnitude of the velocity is called the speed. The direction of the velocity is the direction of motion.

If **r** is the position vector of an object then the velocity **v** is given by

$$\mathbf{v} = \frac{d\mathbf{r}}{dt}$$

The units of velocity are meters per second, written m s⁻¹.

If $\mathbf{r}_1$ and $\mathbf{r}_2$ are the position vectors of an object at two times t_1 and t_2, respectively, then the average velocity of the object is:

$$\text{average velocity} = \frac{\mathbf{r}_2 - \mathbf{r}_1}{t_2 - t_1}$$

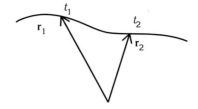

Example 1:

An object moves so that its position vector is given in *Cartesian coordinates* by:

$$\mathbf{r} = 4 \cos (\omega t)\mathbf{i} + 4 \sin (\omega t)\mathbf{j}$$

where ω is a constant and t is the time. Find the velocity of the object and describe its path.

Solution:

Differentiating the position vector gives

$$\mathbf{v} = \frac{d\mathbf{r}}{dt} = -4\omega \sin (\omega t)\mathbf{i} + 4\omega \cos (\omega t)\mathbf{j}$$

The magnitude of the position vector is given by:

$$|\mathbf{r}| = \sqrt{[4 \cos (\omega t)]^2 + [4 \sin (\omega t)]^2} = 4$$

so the path of the object is a circle of radius 4 units.

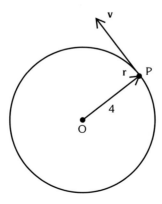

The figure above shows the path of the object and the velocity vector at some point P. The magnitude of the velocity is 4ω and has direction along the tangent to the circle at P.

Example 2:

A cyclist rides with constant speed along a winding road. Draw arrows to represent the velocity vector at the points shown. Is the velocity constant at any time?

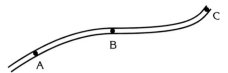

Solution:

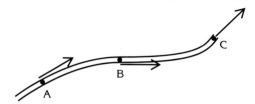

The arrows all have the same length because the speed is constant. The direction of the arrow is a tangent at each point.

The velocity is not constant because the direction of the path, and hence the velocity vector, is always changing.

velocity–time graphs: consider an object moving along a straight line. If $x(t)$ is the position of the object as a function of time t, then the velocity is given by:

$$v = \frac{dx}{dt}$$

If $v > 0$ then x is increasing as t increases, so the object is moving to the right; if $v < 0$ then x is decreasing as t increases, so the object is moving to the left.

We can represent the velocity of the object by drawing a graph of v against t. This is called a velocity–time graph.

Example 1:

The diagram shows the velocity–time graph for an object moving along a straight line.

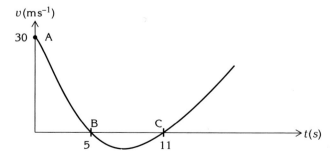

Describe the motion of the object for the first 12 seconds.

Solution:

From the diagram the object starts with speed 30 m s^{-1} moving to the right (i.e. in the direction of x increasing).

After 5 seconds the velocity is zero: the object comes to instantaneous rest.

Between B and C the velocity is negative, so the object is now moving to the left (i.e. in the direction of x decreasing).

After 11 seconds the object comes to instantaneous rest again.

After this time the object moves to the right, since $v > 0$.

The following figure shows the path of the object.

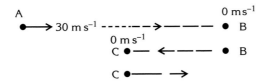

The slope of the velocity–time graph is the acceleration of the object.

The integral of the velocity–time graph is the position of the object.

The area between the velocity–time graph and the t-axis is the distance traveled.

Example 2:

A cyclist travels along a straight road; the cyclist's velocity–time graph is shown in the figure below.

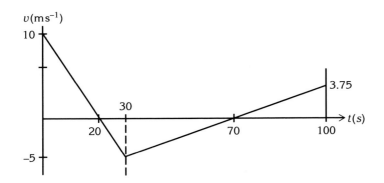

(a) Find the acceleration of the cyclist.
(b) Find the total distance traveled by the cyclist during the first 100 seconds.
(c) If the initial position of the cyclist is $x = 0$, find the position of the cyclist after 100 seconds.

Solution:

(a) For the first 30 seconds the acceleration is $-15/30 = -0.5$ m s^{-2}; during the next 70 seconds the acceleration is $8.75/70 = 0.125$ m s^{-2}.

(b) The total area under the graph is

$$(\tfrac{1}{2} \times 20 \times 10) + (\tfrac{1}{2} \times 50 \times 5) + (\tfrac{1}{2} \times 30 \times 3.75) = 281.25 \text{ meters}$$

(c) The integral of the velocity–time graph is

$$(\tfrac{1}{2} \times 20 \times 10) - (\tfrac{1}{2} \times 50 \times 5) + (\tfrac{1}{2} \times 30 \times 3.75) = 31.25$$

The position of the cyclist after 100 seconds is $x = 31.25$

Venn diagram: a Venn diagram is a way of displaying the *events* of an *experiment*.

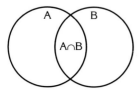

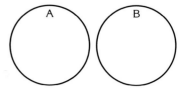

A venn diagram showing two events A and B and their intersection

A venn diagram showing two mutually exclusive events A and B

vertex: a vertex of a two-dimensional shape is a point where two sides intersect. For example, a triangle has three vertices.

A vertex of a three-dimensional shape is a point where three edges intersect. For example a cube has eight vertices.

vertex (node): a point on a *graph* representing a point where *edges* meet.

volume of revolution: the volume formed when a region is rotated through 360° around an axis. The volume formed by rotating the region enclosed by the x-axis, the lines $x = a$ and $x = b$ and the curve $y = f(x)$ through 360° around the x-axis is given by:

$$\int_{a}^{b} \pi y^2 \, dx$$

Example:

The region enclosed by the x-axis, the lines $x = 1$, $x = 4$ and the curve $y = x^2$, is rotated through 360° around the x-axis. Find the volume of the solid that is formed.

Solution:

The diagram shows the region that is rotated and the solid that is formed.

The volume of the solid can be found using

$$V = \int_{a}^{b} \pi y^2 \, dx$$

In this case this gives:

$$V = \int_{1}^{4} \pi (x^2)^2 \, dx$$

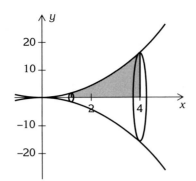

$$= \int_1^4 x^4 \, dx$$

$$= \pi \left[\frac{x^5}{5} \right]_1^4$$

$$= \pi \left(\frac{4^5}{5} - \frac{1}{5} \right)$$

$$= \frac{1023\pi}{5}$$

watt: The watt (W) is the SI unit of power and is named after the Scottish engineer James Watt (1736–1789)

> 1 watt (W) = 1 joule per second (J s⁻¹)

weight: the value on the *edge* of a *network*.

weight: The weight of a body is the *force of gravity* acting on the body due to the gravitational influence of the Earth. The weight of a body varies slightly from place to place on the surface of the Earth. For a mass of 1 kg the standard acceleration due to gravity is 9.81 m s⁻², so the gravitational force on the body due to the Earth has magnitude 9.81 N. The weight of an object of mass 1 kg is taken as 9.81 N. (See also *mass*.)

weighted mean: if the numbers $x_1, x_2, \ldots x_n$ are given the weights $w_1, w_2, \ldots w_n$ then the weighted mean, $\bar{x}_w$, is given by

$$\bar{x}_w = \frac{w_1 x_1 + w_2 x_2 + \ldots + w_n x_n}{w_1 + w_2 + \ldots + w_n} = \frac{\Sigma w_i x_i}{\Sigma w_i}$$

Example:

A student's marks for his coursework are 56%, 62%, 65% and 59%. The pieces of coursework are weighted in the ratio 1:2:3:4 respectively. Calculate the student's average coursework mark.

Solution:

$$\bar{x}_w = \frac{1 \times 56 + 2 \times 62 + 3 \times 65 + 4 \times 59}{1 + 2 + 3 + 4} = \frac{611}{10} = 61.1\%$$

work: the work done when a constant force of magnitude F acting on a body causes the body to move a distance s is defined by the magnitude of the force times the distance when the body moves along the line of action of the force. Work is a *scalar* quantity.

The units of work are joules (J) so that 1 J = 1 N m

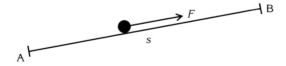

> work done = $F \times s$

For example, if a person lifts an object of mass 20 kg through a height of 3 meters, then the work done by the person is $20g \times 3 = 60g = 588.6$ joules.

If the constant force of magnitude F is directed at an angle θ to the direction of the line AB then

work done $= F\cos\theta \times s$

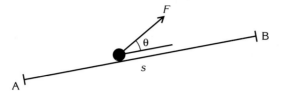

Because of the way that work is defined, we can use the scalar product in the definition:

work $= \mathbf{F.d}$

where $\mathbf{d}$ is the displacement along A to B.

work–energy equation: this is a relationship between the change in *kinetic energy* and the *work* done by a force.

work done $=$ increase in kinetic energy

The equation can be derived by integrating Newton's second law of motion (see *Newton's laws of motion*). For example, if we consider the motion of an object of mass m in one dimension and subject to a constant force F, then the equation of motion of the object is:

$$F = ma = mv\frac{dv}{dx}$$

If we integrate both sides with respect to x and assume that in moving a distance d the speed increases from u to v then:

$$\int F\,dx = \int mv\frac{dv}{dx}$$

$$Fd = \tfrac{1}{2}mv^2 - \tfrac{1}{2}mu^2$$

The left-hand side of this equation is the work done by the force F and the right-hand side is the change in kinetic energy.

Example:

A loaded supermarket cart has a total mass of 80 kg and is at the top of a slope in a parking lot. The cart begins to roll (in a straight line!) down the slope and after 20 meters has a speed of 6 m s^{-1}. The slope of the parking lot is 10°.

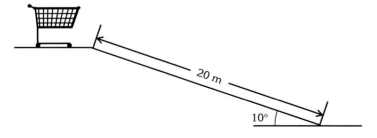

(a) Find the work done by the force of gravity as the cart rolls down the slope.

(b) If the resistance forces are assumed constant, use the work–energy equation to find their magnitude.

(c) At the bottom of the 20 meter slope the parking lot is level. Find how far the cart travels before coming to rest.

Solution:

(a) The force of gravity has magnitude $80g = 80 \times 9.8 = 784$ N and during the motion down the slope the cart drops through $20 \sin (10°) = 3.473$ meters.

work done by the force of gravity $= 784 \times 3.473 = 2722.8$ joules

(b) Let the resistance forces have magnitude R, then the work done by R is $-R \times 20$ joules. (The work done is negative because the force R is pointing up the slope and opposes the motion.)

total work done $= 2722.8 - 20R$

change in kinetic energy $= \frac{1}{2} \times 80 \times 6^2 - \frac{1}{2} \times 80 \times 0^2 = 1440$ joules

Using the work–energy equation:

$$2722.8 - 20R = 1440$$

Solving for R: $R = 64.14$ N

(c) On the level surface of the parking lot the only force acting is the resistance force. Suppose that the cart comes to rest after traveling a further distance d meters:

The work–energy equation gives: $-R \times d = \frac{1}{2} \times 80 \times 0^2 - \frac{1}{2} \times 80 \times 6^2$

Substituting for $R = 64.14$ N and solving for d we get $d = 22.5$ meters

Yates' continuity correction: in testing the independence of two factors in a *contingency table* with a *chi squared* (χ^2) *test*, Yates' continuity correction needs to be applied if there is only one *degree of freedom*. This will occur in a 2×2 contingency table.

The correction gives:

$$\chi^2 = \sum \frac{(|O - E| - 0.5)^2}{E}$$

Example:

A survey of 1250 passengers on long-distance rail journeys gave the following information for numbers of passengers who bought different tickets.

Gender	Class of ticket	
	First	Coach
Male	95	405
Female	165	585

Is there any association between gender and class of ticket?

Solution:

H_0 : gender and class of ticket are independent

H_1 : gender and class of ticket are dependent

Under H_0 the expected frequencies would be

Gender	Class of ticket	
	First	Coach
Male	104	396
Female	156	594

(See *contingency tables* for details.)

The degrees of freedom for a 2×2 contingency table are 1, and so Yates' continuity correction needs to be applied.

$$\chi^2 = \sum \frac{(|O - E| - 0.5)^2}{E}$$

$$= \frac{(|95 - 104| - 0.5)^2}{104} + \frac{(|405 - 396| - 0.5)^2}{396}$$

$$+ \frac{(|165 - 156| - 0.5)^2}{156} + \frac{(|585 - 594| - 0.5)^2}{594}$$

$$= 1.462$$

The 5% χ^2 value with 1 degree of freedom is 3.841 from tables. Since the test statistic is less that the table value, we cannot reject the null hypothesis. Therefore there is no evidence of a link between gender and class of ticket.

zero index: any nonzero number raised to the power 0 is equal to 1. For example $2^0 = 1$.

zero vector: the vector with magnitude zero. No direction is specified for this vector. It is denoted by **0**.